Spreadsheet Applications in Engineering Economics

Alfred L. Kahl
University of Ottawa

William F. Rentz
University of Ottawa

West Publishing Company

St. Paul ■ New York ■ Los Angeles ■ San Francisco

Dedication

To my grandson Andrew William, who has brought new joy to my life. WFR

WEST'S COMMITMENT TO THE ENVIRONMENT

In 1906, West Publishing Company began recycling materials left over from the production of books. This began a tradition of efficient and responsible use of resources. Today, up to 95% of our legal books and 70% of our college texts are printed on recycled, acid-free stock. West also recycles nearly 22 million pounds of scrap paper annually—the equivalent of 181,717 trees. Since the 1960s, West has devised ways to capture and recycle waste inks, solvents, oils, and vapors created in the printing process. We also recycle plastics of all kinds, wood, glass, corrugated cardboard, and batteries, and have eliminated the use of styrofoam book packaging. We at West are proud of the longevity and the scope of our commitment to our environment.

Production, Prepress, Printing and Binding by West Publishing Company.

COPYRIGHT © 1992 by   WEST PUBLISHING CO.
610 Opperman Drive
P.O. Box 64526
St. Paul, MN 55164–0526

ISBN 0–314–87680-4

Note to the User

This book can be used with any of the standard textbooks in engineering economics.

Users of the worksheet template diskette which accompanies this book will need access to an appropriate personal computer with its DOS and electronic spreadsheet software.

The worksheet templates on the accompanying diskettes have been prepared with an IBM compatible personal computer and release 2.2 of Lotus 1-2-3, the most widely used spreadsheet. These diskettes are permanently write protected. In addition, most of the worksheet templates are globally protected. You should make working copies as explained in the Copying Diskettes section of Chapter 1.

However, the worksheet templates have been provided in the lowest common denominator format, known as WKS, since almost all the other spreadsheets can read such files. To further enhance compatibility, we have avoided using macros. Thus, you can use the templates with Lotus 1-2-3, Excel, Quattro and Quattro Pro, SuperCalc, VP Planner, and many others.

Whenever this book mentions microcomputer commands, the text will show key cap symbols. Thus, for example, if you are to type the words ENGINEERING ECONOMICS and then depress the enter key, the text will appear as
E N G I N E E R I N G space bar E C O N O M I C S Enter .

Worksheet Templates

Contents

Preface

Virtually every organization in the world today uses personal computers, mostly IBM PCs or compatibles. Furthermore, most of these organizations also use powerful spreadsheet software, such as Lotus 1-2-3. In fact, both the IBM PC and Lotus 1-2-3 are the de facto standards. This book is devoted to generic engineering economics spreadsheet applications on IBM and compatible personal computers.

One of the main reasons for the popularity of personal computers was the availability of spreadsheet software, which was not available in 1981 on mainframe computers. Spreadsheet software is extremely useful to managers. In particular, managers are increasingly performing financial analyses with electronic spreadsheets.

Since all engineering economics proposals must ultimately be approved by the top management of the firm, it is very important for engineers to communicate with managers in a way that managers can easily understand.

Spreadsheet Applications in Engineering Economics seeks to provide future engineers with a basic grounding in the use of spreadsheet software.

Purpose of the Book

The purpose of this supplementary textbook is fourfold:

The primary purpose is to demonstrate how spreadsheet software and worksheet templates are useful in solving engineering economy problems.

The secondary purpose is to teach the reader how to develop worksheet templates.

The tertiary purpose is to provide brief summaries of the key areas which are strong in traditional engineering economy texts.

The final purpose is to fill in many of the gaps in important areas that are weak in traditional engineering economy texts. We concluded that this was necessary after repeated urgings by colleagues and reviewers from around the world. For example, we have included in the Appendix to Chapter 3 a discussion of the testing of the significance of the overall regression equation using the F test. This topic is normally omitted by the leading engineering economy books, but it's essential for practicing professionals. Another important topic, financial data analysis, is covered in Chapter 6.

Organization of the Book

Chapter 1 describes how to use microcomputers and spreadsheets to construct worksheets as well as how to save, print, and retrieve worksheets.

Chapter 2 illustrates the fundamentals of time value of money calculations with some simple worksheets that can be used as building blocks to perform more complex calculations. Present and future worth calculations are presented. An integrated worksheet for the seven basic factors commonly used in engineering economics calculations is also included. Nominal and effective interest rates are discussed. The chapter ends with loan amortization and mortgage worksheets.

Chapter 3 discusses the use of spreadsheets in forecasting net cash flows for engineering economics projects as well as such topics as inflation, depreciation, taxes, and working capital. The appendix to this chapter discusses regression analysis.

Chapter 4 discusses cash flow calculations, including the following topics: weighted average cost of capital (WACC), minimum acceptable rate of return (MARR), discounted cash flow analysis (DCFA), internal rate of return (IRR), net present worth (NPW) comparisons, and equivalent annual worth (EAW) comparisons.

Chapter 5 discusses advanced discounted cash flow topics, including net present worth with taxes; sensitivity, risk, and inflation analyses; as well as replacement analyses. The appendix to this chapter presents a case study of the estimated cash flows of an actual project which has been disguised at the request of the company. The appendix also provides a comparison of the U.S. and Canadian tax systems.

Chapter 6 discusses financial data analysis and ratio analysis. This chapter includes a sample financial data analysis of an actual firm.

Appendix A provides a comprehensive discussion of the most important spreadsheet commands for engineering economics applications.

Appendix B discusses the most important spreadsheet functions for engineering economics applications.

Appendix C provides an introductory discussion of spreadsheet macros.

Acknowledgements

We want to thank those who have helped in the production of this book. Several people reviewed the manuscript and their suggestions provided guidance for major improvements. They include Neil Aukland, New Mexico State University; Andy R. Bazar, California State University, Fullerton; Kenneth E. Case, Oklahoma State University; Steven C. Chiesa, Santa Clara University; S. Deivanaayagam, Tennessee Technological University; Jane M. Fraser, Ohio State University; Scott C. Iverson, University of Washington; Bruce N. Janson, University of Colorado at Denver; Jack R. Lohman, National Science Foundation; Daniel P. Loucks, Cornell University; John W. Lucey, University of Notre Dame; William C. Moor, Arizona State University; Gerald J. Thuesen, Georgia Institute of Technology; and Robert O. Warrington, Louisiana Tech University. In addition, we want to thank several people at West Publishing Company. They include T. Michael Slaughter, Dean W. DeChambeau, Beth Hatton, Ann Swift, and Lori Zurn.

Alfred L. Kahl and William F. Rentz
University of Ottawa

Chapter 1

SPREADSHEET TUTORIAL

This chapter provides a concise introduction to IBM-compatible microcomputers and electronic spreadsheets for those who are not familiar with them. It describes how to operate an IBM-compatible personal computer, how to access and use spreadsheet software, how to construct a worksheet, and how to save, print, and retrieve a worksheet. Although it is based on the most popular spreadsheet, Lotus 1-2-3, its contents are sufficiently generic to be useful on almost all spreadsheet software. If you are an experienced user of spreadsheets (or a Macintosh user), you may want to go to Chapter 2.

Microcomputers

A personal computer is actually a modular system. It is usually composed of four major components: a monitor, a system unit, a keyboard, and a printer.

Figure 1-1 shows a typical IBM type personal computer system, but without a printer. The monitor is on top of the system unit. The keyboard is in front of the system unit.

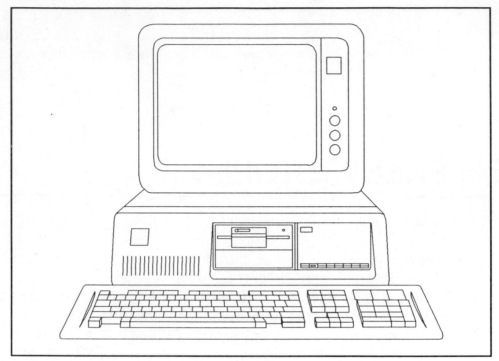

Figure 1-1 An IBM PC Microcomputer System

Monitor

The *monitor* (computer screen) provides information to the user of the computer. After the computer is turned on, each time a key is depressed a signal is sent to the memory and to the screen. The CPU (*Central Processing Unit*) takes information from memory, transforms it according to the program instructions, and sends the results to the screen plus another location in memory. When the computer is turned off, everything in RAM (*Random Access Memory*) is lost, so some form of permanent storage is necessary.

System Unit

Within the recessed area on the right side of the box containing the *system unit*, there is a floppy disk drive on the left and a hard disk drive on the right. The vertical lines on the left of the box are the ventilation slots which allow cooling air to enter. There is an exhaust fan at the rear of the box. The system unit on-off switch is usually in the back of the box on the right or on the right side of the box near the back corner.

Keyboard

The *keyboard* is usually attached to the system unit by a long coiled cord, which allows the user to find a comfortable position. Microcomputer keyboards resemble typewriter keyboards, but they have some additional keys as well.

The *function keys* [F1] through [F12] are arranged along the top of the keyboard in Figure 1-1. The function keys allow one keystroke to perform some action that would otherwise take more keystrokes. On some keyboards, the function keys are in a *function keypad*, containing keys [F1] through [F10], located on the left-hand side of the keyboard.

The *numeric keypad* is the group of keys on the right-hand side of the keyboard. The keys [0] through [9] are used to enter digits when the [Num Lock] key is on.

The keys [2], [4], [6], and [8] also have arrows [↓], [←], [→], and [↑], respectively, on them. If the [Num Lock] key is **not** on, these keys, and the keys [1], [3], [7], and [9] (which also have [End], [PgDn], [Home], and [PgUp], respectively, on them) can be used to move the cursor around the screen. Worksheet navigation is discussed later in this chapter.

Enhanced keyboards (such as the one in Figure 1-1) also have separate [↑], [←], [↓], [→], [Home], [End], [PgUp], and [PgDn] dedicated cursor keys between the numeric keypad and the main keyboard. Thus, the [Num Lock] key can be left always on if these dedicated cursor keys are used. This capability makes spreadsheet software easier to use, as it avoids having to toggle the [Num Lock] key on to enter a number and off to move around the worksheet.

Also between the numeric keypad and the main keyboard are the dedicated [Insert] and [Delete] keys, which duplicate the [Ins] and [Del] functions of the [0] and [.] keys of the numeric keypad. The [Insert] key toggles the insert mode on or off. When it's on, new characters will be inserted at the cursor position; otherwise, the new characters will write over what is already there. The [Insert] key is useful for editing. The [Del] key is also useful for editing. It deletes (or erases) a character at the cursor position. The [BackSpace] key, which is usually on the main keyboard, can also be used to erase a character. However, it erases a one space to the left (backwards) of the cursor position.

The main (typewriter) keyboard contains some additional, non-typewriter keys. The most important of these other keys are the *Control* and *Alternate* keys, labelled (Ctrl) and (Alt). The (Ctrl) and (Alt) keys are often used with other keys to increase the number of functions, without adding unnecessarily to the total number of keys.

The three keys (Ctrl), (Alt), and (Del) also have a special function. If all three of them are pressed at the same time, the computer will do what is known as a *warm boot*. That is, it will restart the machine, but without the time-consuming power-on self-test (or POST) which is described later.

One of the most important keys is the (Enter) (or (Return) or (◄─┘)) key, which is just to the right of the typewriter keypad. When this key is pressed, it signals the computer that it should now perform the command that has just been typed in. Since this key is so important, the enhanced keyboards have a second (Enter) key on the extreme right-hand side of the numeric keypad.

The key labelled (Esc) is the *Escape* key. It is usually found on the upper left corner of the keyboard. It's used to back out of or undo something. So, this key is also a very important key.

The (Shift) shift-arrow keys on the left and right sides of the typewriter part of the keyboard are the shift keys for changing between upper case (capital letters) and lower case (small letters).

The long bar at the bottom (center) of the keyboard is the (space bar) key. On most computers it is blank. In this book it's represented by (space bar) .

Microcomputer Operations

To use the worksheet template diskette which accompanies this book you will need DOS and a spreadsheet, such as Lotus 1-2-3. If spreadsheet software is not on your hard disk (or if you have no hard disk), you will need to use the appropriate spreadsheet software diskettes to load the spreadsheet into your computer before you try to use the templates.

Operating Systems

Every computer needs to have an *operating system* that controls the operation of the machine and the application programs. For most microcomputers, the operating system is called DOS (*Disk Operating System*). DOS is a group of programs normally provided by the maker of the computer. Usually, most of DOS resides on a hard or floppy disk, but on some computers it is in ROM (*Read Only Memory*).

IBM hired another firm, Microsoft, to make the DOS for the IBM personal computers, and it is called PC DOS. Microsoft also wrote MS (for MicroSoft) DOS for the compatibles and clones. The two DOS's are almost the same, except that PC DOS has more of its code in ROM than MS DOS.

Starting Hard Drive Computers

To start the computer **if it has a hard disk**, turn on the monitor and the system unit. This start up procedure is called a *cold boot*. When the computer starts, it performs the *power-on self-test* (POST) to check all the memory chips to see if they are O.K. This POST may take a minute or more. When the computer is ready for you to give it a command, it will display a *prompt symbol*, which is usually the greater than sign **>**.

Starting Floppy Drive Computers

To start the computer **if it does not have a hard disk**, put a DOS diskette into the first floppy disk drive (drive A:) and turn on the monitor and the system unit. This start up procedure is called a *cold boot*. When the computer starts, it performs the *power-on self-test* (POST) to check all the memory chips to see if they are O.K. This POST may take a minute or more. When the computer is ready for you to give it a command, it will display a *prompt symbol*, which is usually the greater than sign **>**.

Floppy diskettes come in 5.25" and 3.5" sizes. Insert the 5.25" size diskette into the horizontal 5.25" drive slot with the rectangular cutout, or notch, on the left when the diskette label is facing up and close the latch. (Some 5.25" diskettes, such as the permanently write protected template diskettes that accompany this book, have no notch. Compare the notchless diskette to a notched diskette to be sure that you insert it properly.) Insert the 3.5" size diskette into the horizontal 3.5" drive slot with the imprinted arrow side up and the arrow pointing into the drive. (Some 3.5" diskette drives are mounted vertically.)

Formatting Diskettes

You will probably be using floppy diskettes for storage. All diskettes must be formatted before information can be stored on them. Thus, your spreadsheet diskettes and the template diskettes accompanying this book are already formatted. **Do not attempt to format these diskettes again or you will lose the valuable information stored on them.** If you haven't already done so, you will, however, need to format some blank diskettes before proceeding with this tutorial. Diskettes are usually formatted with the DOS *format command*. To find out how to do this on your own machine, you should consult the DOS Manual for your computer.

Copying Diskettes

Some computers have two floppy drives. However, even if your computer has only one physical disk drive, it treats this one physical drive as two logical drives. This allows you to copy from one diskette to another. These two logical drives are identified as drives A: and B:. If your microcomputer also has a hard disk, it will normally be drive C:. Even though the hard disk can be used for "permanent" storage, there is still the possibility of its being damaged.

Thus, you should get in the habit of making *backup copies* on your own diskettes of everything you do with the computer. You should have at least two floppies for your data and two more for your copies of the worksheet templates from this book.

You should make a working copy of the template diskettes that accompany this book. The DOS *copy command* can be used to copy data from one diskette to another formatted diskette. To copy the worksheet templates from a diskette in drive A: to a formatted diskette in drive B:, make sure the computer is ready for you to give it a command (the DOS prompt should appear on the video monitor). Then type the following command, which will cause the computer to copy all the files from the disk in drive A: to the disk in drive B: C O P Y space bar A : * . * space bar B : ⏎ .

If you only want to copy one worksheet file, just use its name instead of the wildcard symbol, * . * . Suppose you want to copy the file MYFILE.WKS to the diskette in drive B:. In this example, MYFILE is the file name and .WKS is the extension. (The worksheet template diskettes that accompany this book contain files with the extension .WKS). Type Ⓒ Ⓞ Ⓟ Ⓨ [space bar] Ⓐ : Ⓜ Ⓨ Ⓕ Ⓘ Ⓛ Ⓔ . Ⓦ Ⓚ Ⓢ [space bar] Ⓑ : ⏎ .

Electronic Spreadsheets

Electronic spreadsheet software replaces columnar paper, pencils, erasers, and calculators in performing repetitive calculations. Electronic spreadsheet software provides a worksheet composed of a flexible matrix of columns and rows which are identified by letters and numbers. Electronic worksheets are ideal for answering "what if" questions as well as for performing calculations.

Starting a Working Session

If your computer is not already turned on, follow the steps in the relevant section entitled Starting Floppy Drive Computers or Starting Hard Drive Computers. Make sure the DOS prompt **>** appears before proceeding.

If you have a floppy drive computer, remove the DOS disk from the drive and insert the appropriate spreadsheet software diskette. Then give the appropriate command. For Lotus 1-2-3, this would be Ⓐ : ① ② ③ ⏎ .

If you have a hard disk computer your spreadsheet software is probably already on the drive. If you have a batch file, as most hard disk users do, then merely type it in.

Otherwise, you must first change to the appropriate directory. Use the **C**hange **D**irectory command, followed by the name of the directory (for example: **\123**, for Lotus 1-2-3), and then the software's name. Ⓒ Ⓓ \ ① ② ③ ⏎ ① ② ③ ⏎ .

The Blank Worksheet

The *blank worksheet*, as shown in Figure 1-2, should now appear on your computer screen. When your computer screen looks like the figure, you are ready to begin working. The screen is able to show 25 lines of 80 characters each. The first three lines are used by the control panel, which is explained in the next section. The highlighted border area across line 4 and down the left side of the spreadsheet for the next twenty lines identifies the columns and rows of the spreadsheet matrix. There are eight columns, A through H, and twenty rows on the screen. (This is actually only the upper left corner of the worksheet, but it is all that will fit on the screen.) The intersection of each column and row is called a *cell*. Each cell can be uniquely identified by a combination of letters and numbers. The first cell in the upper left corner is A1. The next cell on the diagonal is B2, then C3, etc.

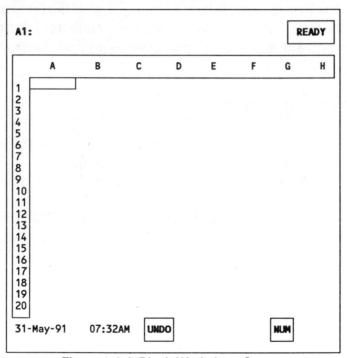

Figure 1-2 A Blank Worksheet Screen

Status Indicators

The last line on the screen is used by the *status indicators*.

When UNDO appears, it means the undo feature is available by simultaneously pressing the Alt F4 keys. The undo feature restores whatever worksheet data and setting that existed when the worksheet was last in the READY mode.

CIRC indicates that the worksheet contains a formula that refers to itself (a circular reference).

When CALC appears, it means that the spreadsheet's formulae need to be recalculated. Press the `F9` key.

OVR indicates that the `Ins` key has been pressed and overstriking is on. Instead of inserting the character you type to the left of the cursor, the typed character replaces the existing character at the cursor position. Although Lotus calls this overstriking, some other software calls this typeover.

NUM indicates that the *Num Lock* is on. This signifies that you can use the numeric keypad to enter numbers.

CAPS indicates that the *Caps Lock* is on. If you type on the alphabetic keys, capital letters will be entered.

The Control Panel

The *control panel*, which occupies three lines of a worksheet screen, contains information about the current cell, mode, and commands. It may be either at the top or bottom of the worksheet. An example control panel is presented in Figure 1-3.

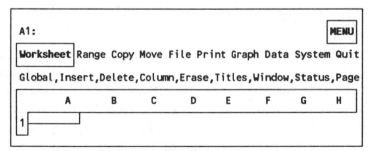

Figure 1-3 The Control Panel

The first line of the control panel shows the current cell address, cell contents, column width, format, and protection status. The second line serves as the prompt for the spreadsheet software. The third line of the control panel displays either a submenu or a description of the command currently highlighted on the menu line above it. Thus, the third line of Figure 1-3 displays the submenu for the highlighted **W**orksheet command.

When the cursor is on a cell, the cursor is called the *cell pointer*. The *cell address* always appears on the control panel. In Figure 1-3 the cell address is shown as A1. The other items, which follow the colon, are only shown after they have been established.

The second line of the control panel (which is blank in Figure 1-2) displays the current entry when you are in the process of creating or editing it. This line displays the main menu of slash commands whenever you press the *slash* (/) key in the ready mode. This is the signal to the spreadsheet that you are going to give it a command. Figure 1-3 shows the control panel after you press (/). Note that the **W**orksheet command is highlighted on line 2.

Mode Indicator

The *mode indicator* is at the upper right hand corner of the command panel. When you start the spreadsheet software, the mode indicator will say READY, as shown in Figure 1-2. It will frequently change to WAIT while some operation is taking place. You will probably see HELP and ERROR often. There are several other indicators that may appear. They are all self-explanatory, like the MENU mode illustrated in Figure 1-3.

Worksheet Structure

The maximum size of the worksheet that will fit in a computer is limited by both hardware and software. The hardware limit depends on the amount of memory that's available for loading the spreadsheet. The software limit depends on both the brand and the version of the spreadsheet you use. For example, the maximum worksheet size of Lotus 1-2-3 versions (releases) varies from 64 to 256 columns and from 256 to 8196 rows. The smaller dimensions are found in the student versions, while the larger dimensions are found in the software intended for corporate use. In most spreadsheets, each of the columns is initially 9 characters wide. Thus, the screen will show you a worksheet matrix that is 72 (i.e. 8 columns) characters wide by 20 rows down. If you need to see more, use the cursor keys, (→) or (↓) to scroll to the right or down, as necessary.

Worksheet Ranges

A *range* is any rectangular block of cells, defined by the two diagonally opposite cells at its extremes, separated by at least one period. In Figure 1-4, *Alpha* is the name of the range B3.C3, *Beta* identifies the range B7.B11, and *Gamma* identifies D6.G11. You can name a range by typing Ⓐ Ⓡ Ⓝ Ⓒ , it's name, such as Ⓐ Ⓛ Ⓟ Ⓗ Ⓐ ⏎ , and it's limits, such as Ⓑ Ⓔ . Ⓒ Ⓔ ⏎ .

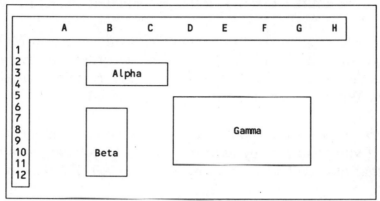

Figure 1-4 Worksheet Ranges

Worksheet Navigation

The *cursor* (or *cell pointer*) indicates where entries will be made in the worksheet. You can make entries whenever the MODE indicator shows READY. You can move the cursor in the following ways:

The *arrow-keys* ← ↑ ↓ → can be used to move the cursor left, up, down, or right one cell at a time. For longer distance moves, the express keys, PgUp and PgDn, can be used. These keys move the cursor up or down one page at a time (i.e. 20 rows).

There are three ways for even longer distance moves. The Home key sends the cursor to the upper left-hand corner (cell A1) of the worksheet. The End key followed by the Home key sends the cursor to the lower right hand corner of the active worksheet. (That is, the last cell which contains an entry.) Depressing the F5 **Go To** key will cause the program to ask you where you want to go. Type in the desired cell address and then press the Enter key. The cursor will immediately move to the requested cell.

If you try to do anything that is not allowed, most spreadsheet software will beep at you. Thus, you will know that the computer won't let you do whatever it was that you tried to do.

Cell Contents

A cell may contain labels, numbers, or formulae. You can enter anything into a cell by typing it. Once a worksheet has been set up, any change in any cell causes the entire worksheet to be recalculated almost instantly.

When you type in the first character of an entry, the MODE indicator will change to either LABEL or VALUE, according to the type of entry being made. As you type, each character will appear on the second line of the control panel. When you are finished, press the (Enter) key.

You can edit as you type if you watch the control panel and use the (BackSpace) key. If you have already used the (Enter) key, you can edit the contents of any cell that already has an entry by using the (F2) *Edit* key.

To erase a cell entry, use the (/)(R)(E) **R**ange **E**rase command.

Each cell can contain labels, formulae, or data entries of up to 240 characters (letters, numbers, symbols, or even blank spaces). However, only those that fit in the cell on the screen will be shown. The rest will be in the computer's memory, and they will be used by the computer if they are part of a number or formula.

Label Entries

A *label entry* can be one or more words which must begin with either a label prefix character or any of the characters which do not indicate the start of a number or a formula. Therefore, labels should begin with either a letter or a prefix.

The prefix apostrophe ' will align the label at the left edge of the cell.

The prefix quote " will align the label at the right edge of the cell.

The prefix caret ^ will center the label within the cell.

The prefix backslash \ will fill the cell with the character that follows the backslash.

For example, ⟨\⟩⟨-⟩ can be used to underline columns of numbers to indicate a summation, while ⟨\⟩⟨=⟩ can be used to double underline a total. The single underline for a summation and double underline for a total are conventions normally used in accounting and finance. So, they are followed by most companies as a matter of policy.

Value Entries

A number *value entry* can begin with any of the following characters: **0 1 2 3 4 5 6 7 8 9 . + - $**. However, a number value generally can not include spaces or commas. (There are some exceptions. Outside the U.S., some spreadsheets are set up to use a comma instead of a decimal point. In the U.S., some spreadsheets will accept commas to separate thousands, millions, etc.) Number value entries may have no more than one decimal point. Number value entries may end with a **%** symbol.

When you enter one of the legal beginning characters in the first position of a cell, the MODE indicator will change from READY to VALUE.

While label cells may be completely filled, numeric values must be one character less than the width of the column. Therefore, if you have 8 character numbers in a column, the column must be at least 9 characters wide. Also keep in mind that totals frequently are larger than the numbers within the columns. Thus, additional space may be required.

Formulae

A *formula* is an instruction to perform calculations. Therefore, it is a special type of VALUE entry. When you type in a formula, the formula appears in the control panel, but the value that results from the operation of the formula appears in the cell.

A formula may begin with any of the characters that are legal for a number entry. In addition, formulae may begin with the (, **@** or **#** symbols. To eliminate confusion, a formula must begin with either a plus **+** sign, a minus **-** sign, a **$** sign, or a left **(** parenthesis when the first part of a formula is a cell address. If you tried to start the formula **B1+B2** with a B, most spreadsheet software would consider it a label. This formula must be written **+B1+B2**, or it must be within parentheses, **(B1+B2)**. Alternatively, you may start the formula with **$B1** (if a mixed reference is appropriate) and Lotus will display it as **+$B1**. Exponentiation is indicated in formulae by the caret **^** symbol.

Formulae can use cell designators, such as **B2**, with arithmetical symbols, such as **+**, to build formulae. A cell, such as B5, for example, might contain the formula **+B2+B3-B4**. The leading arithmetic operator + tells the software that this is a formula. This formula tells the computer to add the values in cells B2 and B3 and then subtract from this result the value in cell B4, putting the end result in the cell B5, which contains the formula. These cell addresses are *relative references*. That is, they really tell the computer to take values from cells that are three, two, and one cells above the one containing the formula. Frequently, you will need to make sure that certain cells, which contain input data for calculations, are used. The **$** symbol added to the cell address **$B$2** indicates an *absolute reference*. Thus, the software must use this particular cell in the calculation of the formula. If you want the row to be constant, while the column varies, use a *mixed reference or address*, such as **D$4**.

Order of Operations

The standard spreadsheet order of operations is to do parentheses first, then exponentiation, then multiply and divide, then add and subtract, working from left to right.

Saving a Worksheet

To save a worksheet, first make sure there is a *formatted* data diskette in the default data drive, usually B: (If your spreadsheet software or DOS diskette should happen to be in the drive, remove it and insert the data disk.) Then type ⧸ Ⓕ Ⓢ for the **F**ile **S**ave menu. (In some spreadsheets, such as SuperCalc, the Ⓕ is not needed.) The **Enter name of file to be saved:** prompt should appear on the command line. (If something follows the prompt, press the ⌁BackSpace⌁ key repeatedly until only the prompt appears. On some spreadsheet software you must type one character before using the ⌁BackSpace⌁ key.) Type in Ⓑ ⦂ Ⓜ Ⓨ Ⓕ Ⓘ Ⓛ Ⓔ ⦁ Ⓦ Ⓚ Ⓢ . The light on the disk drive should light up as the file is saved. (This assumes that MYFILE.WKS is a new file. See Appendix A for information on saving when you wish to replace an existing file.)

It is strongly recommended that you save your worksheets often as you work, and also before printing them. If they do not print the way you want, you can retrieve them and modify them accordingly. If you did not save your work and something goes wrong during printing, then you will have a problem. There is a slogan, frequently used in computing but not unique to it, called Murphy's Law, which says that anything that can go wrong, will go wrong. O'Reilly's corollary of this law is, of course, that whatever will go wrong will happen at the worst possible moment, such as just before your report is due.

Printing a Worksheet

The size of the paper and the type of printer are important considerations at printing time. A standard letter size sheet of paper can hold a maximum of 66 lines of print in the standard line height. Most spreadsheet software reserves six lines for the header and footer (and their spacers), whether you use them or not. However, you can usually change the default top and bottom margins, if you need to do so. It is recommended, as a rule of thumb, that you consider printing no more than 60 lines per page. This is especially true if you are using a laser printer.

The usual default left margin is 4 spaces and the right margin is 76 spaces from the left edge of the paper. Thus, you can print only 72 characters of normal size print on a line. This is the same as what you see on the screen.

To print a worksheet using the defaults, type `/` `P` `P` to get the printer menu, and then type `R` to select **R**ange. At the **Enter print range:** prompt, type in the appropriate range, *viz.*, `A` `1` `.` `B` `1` `7` `Enter` . Press `P` for **P**age, to advance the paper to the top of the next page. (Many laser printers will not print anything unless the **P** command is given.) Press `G` for **G**o and the worksheet will be printed, provided that the printer is turned on, it is connected to the computer (on-line), and it has paper in it. (Some printers may require the **P** command after the **G** command.) Then press `Q` for **Q**uit. This will exit from the printing menu.

Exporting a Worksheet (Printing to Disk)

To export a worksheet, type `/` `P` `F` , then follow the other printing steps above. The worksheet will be printed on the disk as a .PRN file, which can be imported by word processors.

Ending a Working Session

If you are finished with one worksheet and wish to retrieve another, you must first clear the screen, using the ⌿ⓌⒺⓎ command.

If you want to exit the spreadsheet and use some other software, type ⌿Ⓠ↵ . If you are finished with the computer, you may just turn it off.

Retrieving a Worksheet File

The accompanying diskettes contain partially completed template files for most of the worksheets discussed in this book. Most of the templates are globally protected with only some cells unprotected so you can enter data in them. This has been done to prevent you from accidentally destroying the worksheet formulae. However, you can at any time retrieve the template files and modify them, if you first disable the protection. This is done by typing ⌿ⓌⒼⓅⒹ. Global protection is restored by typing ⌿ⓌⒼⓅⒺ .

To retrieve a worksheet file that's been saved to disk, you need a blank worksheet on your screen in the READY mode, like Figure 1-2. Remove any other diskette from drive B: and insert the working copy of the appropriate template diskette. Type ⌿ⒻⓇ for File Retrieve.

If B: is the spreadsheet's default data drive, some of the names of the worksheet files stored on the template diskette will appear on the third line of the control panel. You can use the right arrow → key to highlight the name of the desired worksheet, then press the Enter key to retrieve it.

If drive B: is not the default drive, line 2 of control panel will say **Name of file to retrieve:** followed by something other than **B:**. If this occurs, tap the BackSpace key until only **Name of file to retrieve:** appears on line 2 of the control panel. (You may have to type a character before pressing BackSpace .) For example, if PWF.WKS, is the worksheet you wish to retrieve, type ⒷⒻ︓ⓅⓌⒻ．ⓌⓀⓈEnter . You should then see this worksheet on the screen. PWF.WKS is the first worksheet discussed in the next chapter.

Chapter 2

TIME VALUE OF MONEY

Engineers frequently need to make financial calculations involving the time value of money. No matter where an engineer may work, it will often be necessary to invest money in a new project. Engineering projects usually involve large expenditures in the present or perhaps over a few years for additions to fixed assets (assets with an expected useful life of more than one year) as well as some additional current assets (assets whose life is one year or less) when operation of the new facilities actually begins. These additional assets (or their appropriate time value if outlays are over more than one year) are the *first cost* of the project. Fixed assets can be depreciated. Depreciation is deducted from revenues in calculating taxes. Thus, cash flows are affected. See Chapter 3 for more on these key topics. The first cost of the investment will be followed by a number of future cash inflows. These future cash inflows must be expected to exceed the first cost of the investment if the investment is to be economic.

Discounted cash flow analysis (discussed later) is the method used to evaluate investment projects which require outlays of capital, at one or more points in time, in order to generate future cash flows, at other points in time. Cash flows at different times have different *present worths* (values) because of the time value of money. In particular, a dollar today can be invested to earn interest. Hence, it is more valuable than a dollar to be received some time in the future. Thus, future cash flows must be discounted back to the present so that the various cash flows can be compared on an equivalent basis. Suppose the first cost of the investment involves cash outlays over more than one year. Then it is necessary to calculate the *future worth* (value) of these investment outlays at a time (decision point) just prior to the commencement of the cash inflows.

It is usually necessary to have these investment proposals approved by managers, perhaps several levels higher in the organization. *Spreadsheet software can be very useful in these situations, since it is familiar to managers and used every day by financial people nearly everywhere.*

This chapter illustrates the fundamentals of time value of money calculations with some simple worksheets that can be used as building blocks to perform more complex calculations. We begin with present and future worth calculations and proceed to build an integrated worksheet for all seven of the basic factors: the single payment compound amount (future worth) and single payment present worth factors plus the uniform series sinking fund, uniform series capital recovery, uniform series compound amount, uniform series present worth, and uniform series arithmetic gradient factors. Next, we discuss nominal versus effective interest rates. Finally, we present a worksheet for calculating the amortization of a loan.

Engineers typically speak of the worth of an investment, such as present worth or future worth. Managers, however, call the same things *present value* and *future value.* You should be familiar with both terminologies.

Managers usually use tables of interest factors of one particular type, such as present value, calculated for many different possible interest rates. Engineers, on the other hand, usually use tables of different interest factors for a given interest rate. This chapter presents both types of tables.

Present and Future Worths

Suppose you want to obtain the present worth of a future cash flow. To do this you can multiply the future cash flow by a single payment present worth factor for the relevant interest rate and number of periods. For example, to find the present worth (value) of $10,000 to be received four years from now if the interest rate is 3%, multiply $10,000 by the appropriate factor 0.8885 to obtain $8,885.

Figure 2-1 shows a portion of the 25 period PWF (**Present Worth Factors**) worksheet. The 3%, 4 year factor 0.8885 is shown in cell D9.

Note that cell A1 of Figure 2-1 contains the name of the worksheet, PWF, in this case, and that cell D1 contains the label Z30, which indicates that the lower right hand cell of the worksheet is Z30. We use these two cells to remind us of the name and last cell of the worksheet and recommend that you do the same. Whenever you want to print a worksheet you must tell the computer the range you want to print. With our system, the cell D1 indicates this important information. Throughout this book a worksheet name appears in either cell A1 or B1, and there typically is a label in cell D1 that indicates the lower right hand corner of the worksheet. (Occasionally this label appears further to the right if additional columns are visible.)

The formulae for the PWF worksheet are shown in Figure 2-2.

The PWF worksheet is globally formatted to four decimal places. This was done by typing `/` `W` `G` `F` `F` `4` `⏎` .

Note that an interest rate is entered in row 4 starting with column B and the number of periods is in column A starting with row 5. Thus, the formula in each cell refers to the relevant interest rate and number of periods. The interest rate cells are formatted for percentages with the default value of two decimal places. This was done by typing `/` `R` `F` `P` `⏎` `B` `4` `.` `Z` `4` `⏎` .

	A	B	C	D
1	PWF			Z30
2	PRESENT	WORTH OF $1	DUE AT THE END	OF N PERIODS
4	PERIODS	1.00%	2.00%	3.00%
5	0	1.0000	1.0000	1.0000
6	1	0.9901	0.9804	0.9709
7	2	0.9803	0.9612	0.9426
8	3	0.9706	0.9423	0.9151
9	4	0.9610	0.9238	0.8885

Figure 2-1 PWF Table

	A	B	C	D
4	PERIODS	1.00%	2.00%	3.00%
5	0	1/(1+B$4) ^ $A5	1/(1+C$4) ^ $A5	1/(1+D$4) ^ $A5
6	$A5+1	1/(1+B$4) ^ $A6	1/(1+C$4) ^ $A6	1/(1+D$4) ^ $A6
7	$A6+1	1/(1+B$4) ^ $A7	1/(1+C$4) ^ $A7	1/(1+D$4) ^ $A7
8	$A7+1	1/(1+B$4) ^ $A8	1/(1+C$4) ^ $A8	1/(1+D$4) ^ $A8
9	$A8+1	1/(1+B$4) ^ $A9	1/(1+C$4) ^ $A9	1/(1+D$4) ^ $A9

Figure 2-2 Formulae for the PWF Table

The column A formulae in Figure 2-2 for generating the period numbers work well. Only the formula **$A5+1** in cell A6 needs to be typed. The **C**opy command is used to copy the formula, with appropriate indexing, into cells A7 through A30, for periods 2 through 25. The periods cells are formatted to zero decimal places. This was done by typing ⑦⑧⑨⑩⑪↵Ⓐ⑤.Ⓐ③⓪ ↵.

There is an alternative method for generating the period numbers in cells A5 through A30 by using the **D**ata **F**ill command. Move the cursor to cell A5 and invoke the Data Fill command by typing ⑦ⒹⒻ. The worksheet control panel prompts for the **F**ill range. Type ⦁ and use the ↓ arrow key repeatedly until **A5..A30** appears. Then press the Enter key. In response to the **S**tart prompt, type ⓪ Enter or just Enter if zero appears. In response to the **S**tep prompt, type ① Enter or just Enter if 1 appears. Finally, in response to the **S**top prompt, type ②⑤ Enter and 0 through 25 will appear in cells A5 through A30, respectively.

The **D**ata **F**ill command actually was used to fill the interest rate cells B4 through U4 with the rates 1% through 20%. The remaining percentages in cells V4 through Z4 were individually entered.

Obviously, an analogous FWF (**F**uture **W**orth **F**actors) worksheet could easily be developed for a single payment compound amount by eliminating **1/** from the factor cell formulae in Figure 2-2.

Integrated Tables

In contrast to the tables usually used by managers, engineers typically use tables of several different types of factors for a given interest rate. Figure 2-3 shows such a table. The FACTORS worksheet for this table is included on the diskettes that accompany this book. The key formulae which produced Figure 2-3 are shown in Figures 2-4A through D.

Note that the 40% interest rate is entered in the FACTORS worksheet as either 40% or .4 in cell A1. The cell A1 is formatted as percentage with the default value of two decimal places. Thus, 40% displays as 40.00% in cell A1.

The exponent in each formula refers to the cell at the beginning of the row, which contains the number of periods. The Data Fill command was used to generate the period numbers 1 through 25 in cells A8 through A32.

Row 7 in Figure 2-3 shows the ANSI terminology for each factor.

	A	B	C	D	E	F	G	H
1	40.00%	FACTORS		H32				
3		SINGLE	PAYMENT			UNIFORM	SERIES	
4		Compound	Present	Sinking	Capital	Compound	Present	Arithmetic
5		Amount	Worth	Fund	Recovery	Amount	Worth	Gradient
6		Factor	Factor	Factor	Factor	Factor	Factor	Factor
7	N	(F/P,i,N)	(P/F,i,N)	(A/F,i,N)	(A/P,i,N)	(F/A,i,N)	(P/A,i,N)	(A/G,i,N)
8	1	1.4000	0.71429	1.00000	1.40000	1.0000	0.71429	0.00000
9	2	1.9600	0.51020	0.41667	0.81667	2.4000	1.22449	0.41667
10	3	2.7440	0.36443	0.22936	0.62936	4.3600	1.58892	0.77982
11	4	3.8416	0.26031	0.14077	0.54077	7.1040	1.84923	1.09234

Figure 2-3 40% Interest Factors for Discrete Compounding

	A	B	C
3		SINGLE PAYMENT	SINGLE PAYMENT
7	N	(F/P,i,N)	(P/F,i,N)
8	1	$(1+\$A\$1)\wedge\$A8$	$1/(1+\$A\$1)\wedge\$A8$
9	2	$(1+\$A\$1)\wedge\$A9$	$1/(1+\$A\$1)\wedge\$A9$
10	3	$(1+\$A\$1)\wedge\$A10$	$1/(1+\$A\$1)\wedge\$A10$
11	4	$(1+\$A\$1)\wedge\$A11$	$1/(1+\$A\$1)\wedge\$A11$
32	25	$(1+\$A\$1)\wedge\$A32$	$1/(1+\$A\$1)\wedge\$A32$

Figure 2-4A Formulae for 40% Interest Factors

	A	D	E
3		UNIFORM SERIES	UNIFORM SERIES
7	N	(A/F,i,N)	(A/P,i,N)
8	1	$\$A\$1/((1+\$A\$1)\wedge\$A8-1)$	$\$A\$1/(1-1/((1+\$A\$1)\wedge\$A8))$
9	2	$\$A\$1/((1+\$A\$1)\wedge\$A9-1)$	$\$A\$1/(1-1/((1+\$A\$1)\wedge\$A9))$
10	3	$\$A\$1/((1+\$A\$1)\wedge\$A10-1)$	$\$A\$1/(1-1/((1+\$A\$1)\wedge\$A10))$
11	4	$\$A\$1/((1+\$A\$1)\wedge\$A11-1)$	$\$A\$1/(1-1/((1+\$A\$1)\wedge\$A11))$
32	25	$\$A\$1/((1+\$A\$1)\wedge\$A32-1)$	$\$A\$1/(1-1/((1+\$A\$1)\wedge\$A32))$

Figure 2-4B Formulae for 40% Interest Factors

	A	F	G
3		UNIFORM SERIES	UNIFORM SERIES
7	N	(F/A,i,N)	(P/A,i,N)
8	1	$(((1+\$A\$1)\wedge\$A8)-1)/\$A\$1$	$(1-1/((1+\$A\$1)\wedge\$A8))/\$A\$1$
9	2	$(((1+\$A\$1)\wedge\$A9)-1)/\$A\$1$	$(1-1/((1+\$A\$1)\wedge\$A9))/\$A\$1$
10	3	$(((1+\$A\$1)\wedge\$A10)-1)/\$A\$1$	$(1-1/((1+\$A\$1)\wedge\$A10))/\$A\$1$
11	4	$(((1+\$A\$1)\wedge\$A11)-1)/\$A\$1$	$(1-1/((1+\$A\$1)\wedge\$A11))/\$A\$1$
32	25	$(((1+\$A\$1)\wedge\$A32)-1)/\$A\$1$	$(1-1/((1+\$A\$1)\wedge\$A32))/\$A\$1$

Figure 2-4C Formulae for 40% Interest Factors

	A	H
3		UNIFORM SERIES
7	N	(A/G,i,N)
8	1	1/$A$1-$A8/(((1+$A$1)^$A8)-1)
9	2	1/$A$1-$A9/(((1+$A$1)^$A9)-1)
10	3	1/$A$1-$A10/(((1+$A$1)^$A10)-1)
11	4	1/$A$1-$A11/(((1+$A$1)^$A11)-1)
32	25	1/$A$1-$A32/(((1+$A$1)^$A32)-1)

Figure 2-4D Formulae for 40% Interest Factors

For pedagogical purposes, each cell in Figures 2-4A through D clearly shows the complete expression for calculating the factor. This involves repeatedly calculating (1+$A$1) and 1/(1+$A$1). This is **not** an efficient programming practice.

If you wish to speed up the worksheet, first disable the global protection by typing ⌷W⌷G⌷P⌷D . Then move the cursor to cell B8 and type ⌷1⌷+⌷$⌷A ⌷$⌷1⌷↓ . This replaces the formula (1+$A$1)^$A8 in cell B8 with 1+$A$1 and moves the cursor to cell B9. Now type in ⌷$⌷B⌷$⌷8⌷^⌷$⌷A⌷9⌷Enter . This replaces the formula (1+$A$1)^$A9 in cell B9 with $B$8^$A9 and leaves the cursor on cell B9. Finally, copy the new formula in cell B9 to cells B10 through B32 by typing ⌷/⌷C⌷Enter⌷B⌷1⌷0⌷.⌷B⌷3⌷2⌷Enter .

Since the mixed reference $A9 is used as the exponent in cell B9, the exponent reference is incremented by one row as the formula is copied to successive rows of column B. For example, the formulae in cells B10 and B32, respectively, should now be $B$8^$A10 and $B$8^$A32.

We have suggested improvements to the single payment compound amount factors. Obviously, similar improvements can be made to the other factors. Once you are finished making improvements, you may restore global protection by typing ⌷/⌷W⌷G⌷P⌷E . We suggest that you select a new name such as FASTFACT (fast factors) and type this new name in cell B1. Then follow the instructions in Chapter 1 for saving a worksheet.

Nominal Versus Effective Interest Rates

In North America the interest rate on a loan is stated as a nominal interest rate r per year combined with a number of compounding periods m per year. This nominal interest rate per year is often referred to as the annual percentage rate or APR. The effective compounding period rate i is simply r/m. The effective interest rate per year is $y = (1+i)^m -1$. Thus, the effective interest rate per year is higher than the nominal interest rate if compounding is done more than once per year.

Suppose the number of payment periods p per year differs from the number of compounding periods associated with the loan's nominal interest rate per year. Then an effective payment period interest rate j when compounded p times should equal $1 + y$. Thus, $j = (1+y)^{1/p} -1$. Furthermore, the nominal interest rate per year $s = p * j$ with compounding p times per year is equivalent to the rate r with compounding m times per year in the sense that both generate the same effective interest rate y per year.

Loan Amortization

A loan amortization schedule is frequently needed because tax laws allow certain kinds of interest payments to be deducted in determining taxable income. However, many loan payments include both interest and repayment of the principal of the loan. Since only the interest portion of the loan payment is deductible, it is necessary to be able to find out how much it is.

The AMORTGEN worksheet (Figure 2-5) is a generalized amortization worksheet. The following input data are required for AMORTGEN: the loan amount P (which is a present worth) is entered in cell B7; the nominal interest rate r per year is entered in cell B8; the number of compounding periods m per year is entered in cell B9; the number of payment periods p per year is entered in cell B10; the number of years n to amortize the loan is entered in cell B11; and the loan term t in years is entered in cell B12. The loan term must be less than or equal to the number of years n to amortize the loan.

	B	C	D	E	F
1	AMORTGEN				F391
3	GENERALIZED LOAN AMORTIZATION SCHEDULE				
5	INPUT DATA				
7		= P	= Loan amount		
8		= r	= Nominal interest rate per year		
9		= m	= # of compounding periods per year		
10		= p	= # of payment periods per year		
11		= n	= Amortization in years		
12		= t	= Loan term in years, t < or = n		
14	OUTPUT DATA				
16		= A	= Payment per period		
17		= RB	= Remaining balance at end of term		
18		= PAY	= Total payments during term		
19		= INT	= Total interest paid during term		
20		= PRN	= Total principal paid during term		
21		= i = r / m = Effective compounding period rate			
22		= y	= Effective interest rate per year		
23		= j	= Effective payment period rate		
24		= s = p * j = Equivalent nominal interest rate per year			
25		= N = p * n = # of payment periods to amortize			
26		= T = p * t = # of payment periods during term			
28			Interest	Principal	Remaining
29	Payment Period	Payment	Portion	Portion	Balance
30	--------------	-------	-------	-------	-------

Figure 2-5 Generalized Loan Amortization Schedule Worksheet

There are two important restrictions that must be noted about this worksheet. First, the loan amount must be entered with no more than 7 digits to the left of the decimal point. (That is, loans must be smaller than $10 million if you wish answers to the nearest penny). Second, the total number of payment periods $T = p * t$ must be ≤ 360.

The first restriction is solely to insure that columns B through F are simultaneously visible on the monitor. In particular, the global column width in AMORTGEN is set to 14. You could allow loans of say less than $10 billion by invoking the **W**orksheet **G**lobal **C**olumn-width command by typing `/ W G C 1 8 Enter` . Note that the column width must be increased by 4 for every 3 additional significant digits because commas are displayed in the currency format.

The second restriction was arbitrarily set to be large enough to handle monthly payments on a 30 year term mortgage ($T = p * t = 12 * 30 = 360$). Since more than 360 payment periods is a rare occurrence, this maximum was chosen to conserve space on the templates diskettes. Period 1 payment activity appears in row 32 and period 360 activity is in row 391. Thus, F391 was entered as a label in cell F1 to indicate the lower right hand corner of the worksheet.

A larger number of payment periods than 360 can be accommodated by copying the formulae in cells B391 through F391 into additional cells in their respective columns. Appropriate adjustments must also be made to cells B18 through B20, which sum total payments, interest payments, and principal payments in rows 32 through 391 of columns C through E, respectively.

The larger number of periods would also require changes to column A, which is currently hidden from the user. If you want to see what is in column A, you must invoke the column display command, `/ W C D` . Then A1 is entered as the range and the cursor is moved to column A with the left arrow key, `←` .

Scroll down column A using the down arrow key, `↓` . Note that cells A31 through A391 contain the payment period numbers 0 through 360. These entries were created using the **D**ata **F**ill command, which was discussed earlier in this chapter in the Present and Future Worths section. The **D**ata **F**ill command can be used again to extend the number of payment periods as desired within the limitations of your computing environment. If you are still in column A and wish to hide it again, type `/ W C H Enter` .

Figures 2-6A through E show the formulae for the AMORTGEN worksheet.

For example, in cell B16 of Figure 2-6A note the formula **@IF($B$25>0, @ROUND(@PMT($B$7,$B$23,$B$25),2)," ")**. The @PMT($B$7,$B$23,$B$25) function calculates the periodic loan payment based on the loan amount in cell B7, the effective payment period interest rate contained in cell B23, and the number of payment periods to amortize in cell B25. The @PMT(...) function is embedded in the @ROUND(@PMT(...),2) function, which rounds the periodic loan payment to the nearest penny (2 decimal places). The @ROUND(...) function is embedded in the @IF($B$25>0,@ROUND(...)," ") function.

The @IF function calculates and displays the periodic loan payment in cell B16, if the number of payment periods N to amortize the loan in cell B25 is greater than zero. Otherwise, what is displayed in cell B16 is the same as what is between the double set of quote marks, " ", which is nothing.

	B
14	OUTPUT DATA
16	@IF($B$25>0,@ROUND(@PMT($B$7,$B$23,$B$25),2)," ")
17	@IF($B$12>0,$B$7-$B$20," ")
18	@IF($B$12>0,@SUM($C$32..$C$391)," ")
19	@IF($B$12>0,@SUM($D$32..$D$391)," ")
20	@IF($B$12>0,@SUM($E$32..$E$391)," ")
21	@IF($B$12>0,$B$8/$B$9," ")
22	@IF($B$12>0,(1+$B$21) ^ $B$9-1," ")
23	@IF($B$12>0,(1+$B$22) ^ (1/$B$10)-1," ")
24	@IF($B$12>0,$B$10*$B$23," ")
25	@IF($B$12>0,$B$10*$B$11," ")
26	@IF($B$12>0,$B$10*$B$12," ")
31	@IF($B$12>0,0," ")
32	@IF($A32>$B$26," ",$A32)
33	@IF($A33>$B$26," ",$A33)
391	@IF($A391>$B$26," ",$A391)

Figure 2-6A Formulae for Generalized Loan Amortization Schedule

	C
31	@IF($B$12>0,0," ")
32	@IF($A32>$B$26," ",$B$16+@INT($A32/$B$25)*(-$B$16+@ROUND((1+$B$23)*$F31,2)))
33	@IF($A33>$B$26," ",$B$16+@INT($A33/$B$25)*(-$B$16+@ROUND((1+$B$23)*$F32,2)))
391	@IF($A391>$B$26," ",$B$16+@INT($A391/$B$25)*(-$B$16+@ROUND((1+$B$23)*$F390,2)))

Figure 2-6B Formulae for Generalized Loan Amortization Schedule

	D
31	@IF($B$12>0,0," ")
32	@IF($A32>$B$26," ",@ROUND($B$23*$F31,2))
33	@IF($A33>$B$26," ",@ROUND($B$23*$F32,2))
391	@IF($A391>$B$26," ",@ROUND($B$23*$F390,2))

Figure 2-6C Formulae for Generalized Loan Amortization Schedule

	E
31	@IF($B$12>0,0," ")
32	@IF($A32>$B$26," ",$C32-$D32)
33	@IF($A33>$B$26," ",$C33-$D33)
391	@IF($A391>$B$26," ",$C391-$D391)

Figure 2-6D Formulae for Generalized Loan Amortization Schedule

	F
31	@IF($B$12>0,$B$7," ")
32	@IF($A32>$B$26," ",$F31-$E32)
33	@IF($A33>$B$26," ",$F32-$E33)
391	@IF($A391>$B$26," ",$F390-$E391)

Figure 2-6E Formulae for Generalized Loan Amortization Schedule

The @IF function is liberally used throughout the AMORTGEN worksheet to blank a cell display whenever conditions require this. An alternative approach would be to build and invoke a macro command that generates the relevant number of payment periods for the term loan under consideration.

As stated previously in the Note to the User section on page iii at the start of the book, *we have avoided the use of macros to enhance compatibility with different spreadsheet software.*

The formula **@IF($A\$32>\$B\$26," ",\$B\$16+@INT(\$A\$32/\$B\$25)*(-\$B\$16+@ROUND((1+\$B\$23)*\$F31,2)))** in cell C32 and the analogous formulae in cells C33 through C391 deserve comment. If the current period in cell A32 is greater than the number of periods in the term of the loan in cell B26, then the cell C32 appears blank.

Now let us consider the case $A32 \leq \$B\$26$. Then $A32 \leq \$B25$ because $\$B\$26 \leq \$B25$ (*i.e.* the number of payment periods during the term of the loan must be less than or equal to the number of payment periods to amortize). Since the @INT function returns the greatest integer contained in the calculation $A32 / \$B25$, this function returns zero for $A32 < \$B\$25$ and one for $A32 = \$B\$26 = \$B25$. When zero is returned, the loan payment calculated in cell B16 is shown in cell C32. When one is returned, the final amortizing loan payment @ROUND$((1+\$B\$23)*\$F31,2)$ is displayed in cell C32. For a one-period loan (remember, cell A32 contains 1), this will be the same as the contents of cell B16. In general, however, the final payment to exactly amortize a loan may be slightly different than the preceding payments.

Mortgages

Many firms obtain part of the financing needed to build a plant by borrowing money and pledging the plant as collateral for the loan. This type of loan is called a mortgage.

Engineers, as well as many other individuals, usually own a home sometime during their lives. Few individuals have sufficient funds to buy a home outright. Most finance part of the purchase price with a home mortgage.

Suppose you want to borrow $100,000 to finance a home purchase in the United States. The terms quoted to you are a 12% nominal interest rate per year, monthly compounding, monthly payments, 25 year amortization, and 25 year term.

The input data for this mortgage example are shown in cells B7 through B12 in Figure 2-7. The worksheet name in cell B1 has been changed from AMORTGEN to AMORTAHM (for **AMORT**izing an **A**merican **H**ome **M**ortgage) and cell F1 contains the label F331 (to indicate the lower right hand corner of the worksheet). Cells B16 through B26 contain the main results.

	B	C	D	E	F
1	AMORTAHM				F331
7	$100,000.00	= P	= Loan amount		
8	12.00000000%	= r	= Nominal interest rate per year		
9	12	= m	= # of compounding periods per year		
10	12	= p	= # of payment periods per year		
11	25	= n	= Amortization in years		
12	25	= t	= Loan term in years, t < or = n		
16	$1,053.22	= A	= Payment per period		
17	($0.00)	= RB	= Remaining balance at end of term		
18	$315,973.68	= PAY	= Total payments during term		
19	$215,973.68	= INT	= Total interest paid during term		
20	$100,000.00	= PRN	= Total principal paid during term		
21	1.00000000%	= i = r / m	= Effective compounding period rate		
22	12.68250301%	= y	= Effective interest rate per year		
23	1.00000000%	= j	= Effective payment period rate		
24	12.00000000%	= s = p * j	= Equivalent nominal interest rate per year		
25	300	= N = p * n = # of payment periods to amortize			
26	300	= T = p * t = # of payment periods during term			

Figure 2-7 American Mortgage Loan Amortization Worksheet

Now let us consider an example where the number of compounding periods associated with the stated nominal interest rate per year is different from the number of payment periods per year. Furthermore, let us make the term of the loan less than the amortization period.

An excellent real world example is a mortgage on a home in Canada. Canadian home mortgages state the nominal interest rate per year based on semiannual compounding; whereas, payments are typically made monthly. (This contrasts with the U.S., where the nominal interest rate per year on a mortgage is based on monthly compounding.) Generally, in Canada the term of the mortgage is 5 years or less and the amortization is up to 30 years.

Suppose you wish to borrow $100,000 for a Canadian mortgage with a 12% nominal interest rate per year and semiannual compounding. Payments will be made monthly. The amortization period will be 25 years, but the term will be 5 years.

The input data for this mortgage example are shown in cells B7 through B12 in Figure 2-8. The worksheet name in cell B1 has been changed to AMORTCHM, and cell F1 contains the label F91 to indicate the lower right hand corner of this worksheet. Again, cells B16 through B26 contain the main results.

Comparing the main results of these two examples, we note that the Canadian borrower has a lower monthly payment because the nominal interest per year in Canada is based on semiannual compounding instead of the monthly compounding used in the United States. Thus, the effective interest rates per year and per month are both lower in Canada even though the assumed nominal rates are the same.

Before concluding that the Canadian engineer is leading the good life compared to an American engineer, we must hasten to add the following remarks. The nominal interest rate per year on a Canadian mortgage is typically several percentage points above the U.S. rate. In addition, mortgage interest payments in Canada are **not** tax deductible unless the money was borrowed for the purpose of generating income. Also, the nominal interest rate per year on a Canadian mortgage must be renegotiated at the end of the term, typically five years or less; whereas, the U.S. rate will be guaranteed for the whole amortization period if the term is the same as the amortization.

	B	C	D	E	F
1	AMORTCHM				F91
7	$100,000.00	= P	= Loan amount		
8	12.00000000%	= r	= Nominal interest rate per year		
9	2	= m	= # of compounding periods per year		
10	12	= p	= # of payment periods per year		
11	25	= n	= Amortization in years		
12	5	= t	= Loan term in years, t < or = n		
16	$1,031.90	= A	= Payment per period		
17	$95,460.12	= RB	= Remaining balance at end of term		
18	$61,914.00	= PAY	= Total payments during term		
19	$57,374.12	= INT	= Total interest paid during term		
20	$4,539.88	= PRN	= Total principal paid during term		
21	6.00000000%	= i = r / m = Effective compounding period rate			
22	12.36000000%	= y	= Effective interest rate per year		
23	0.97587942%	= j	= Effective payment period rate		
24	11.71055302%	= s = p * j = Equivalent nominal interest rate per year			
25	300	= N = p * n = # of payment periods to amortize			
26	60	= T = p * t = # of payment periods during term			

Figure 2-8 Canadian Mortgage Loan Amortization Worksheet

Chapter 3
FORECASTING CASH FLOWS

This chapter introduces the topic of estimating and forecasting project cash flows. The engineering economics decision making rules, discussed in Chapters 4 and 5, *all* require cash flow data inputs.

Cash Flows

Estimated cash flow data can be obtained from a variety of sources, such as a firm's accounting records, competitive bids, price quotes, standard costs and times, etc. One method discussed in this chapter is to assume that costs are escalating at a constant rate.

Historical data can be analyzed to determine relationships. Regression analyses (see Appendix 3A) can be used to fit a straight line to historical data so that extrapolations can be made.

Forecasting methods can be used to predict future business conditions and sales prospects.

Cash flows can be categorized as *operating cash flows* and *other cash flows*.

Operating Cash Flows

Operating cash flows arise from the normal operations of the firm. They are the difference between sales revenues and cash expenses, including taxes paid.

The major reasons that operating cash flows can differ from accounting profits are:

(1)   All of the taxes shown in the accounting statements may not have to be paid in the current year. Alternatively, actual taxes paid may exceed the tax shown in the accounting records.

(2)   Sales may be on credit. Hence, the sales figure in the accounting records, which is based on the accrual system of accounting, will not coincide with the firm's cash flows.

(3)   Some of the costs, which are shown in the accounting records and deducted from sales to obtain profits, actually may not have to be paid out in the current year.

(4)   Depreciation shown in accounting statements is not a cash flow although it is a very important operating cost.

Thus, operating cash flows could be larger or smaller than accounting profits for any given year.

Other Cash Flows

Other project cash flows arise from the sale of project assets (such as equipment no longer needed) and from any changes in working capital that are necessary to undertake the project.

Working Capital

The term working capital is subject to some ambiguity. Gross working capital is defined as current assets. *Current assets* include cash, accounts receivable, inventories, and temporary investments. Net working capital is defined as current assets minus current liabilities. *Current liabilities* include accounts payable and other short term debts payable within one year.

Many projects will require the firm to increase its working capital. As the proposed investment will increase the size of the firm, and larger firms usually have more cash, it is very likely that the firm's cash account will have to be increased.

A firm that increases its productive capacity will be producing more, and it will need to increase its raw materials inventories. As the volume of production is increased, work in process inventories usually also increase. Finished goods inventories will normally also be increased. As the new products are sold, accounts receivable will increase.

Thus, an important part of most project analyses is the estimation of the required incremental increase in working capital. This increase in working capital usually occurs at an early stage in the life of the project. At the end of the project's economic life, however, the working capital investment may be recovered. Thus, the cash flows in both the first and last years will be affected.

Depreciation and Taxes

Suppose a firm buys a machine and uses it for five years. The cost of the goods produced by the machine must include a charge for its use; this charge is called *depreciation*.

Depreciation reduces profits as calculated by accountants and reported to investors, but it is not a cash outflow. It is, therefore, termed a "noncash" expense.

Accountants report revenues and expenses attributable to a specific period of time, usually one year. According to generally accepted accounting principles, expenses that benefit many periods, such as the purchase of new equipment, must be depreciated or amortized over the period of time the asset is expected to be used.

Historically, an asset's depreciable life was determined by its estimated useful economic life. That is, it was intended that an asset would be fully depreciated at approximately the same time that it reached the end of its useful economic life. This process is still generally used by accountants for the reports they prepare for the firm's investors.

Suppose a firm buys a machine for $100,000 and expects to use it for 5 years after which it will be scrapped and have no value. What would be the annual depreciation expense? It depends on the method used.

There are a only a few different generally accepted accounting methods for calculating the amount of depreciation expense. The most commonly used methods are straight line, units of production, declining balance, and sum-of-the-years digits.

Straight-Line Depreciation

The *straight-line* depreciation expense equals the cost less salvage value divided by the economic life. Thus it would be $100,000/5 = $20,000 per year. At the end of the five years the entire cost would have been written off.

Units of Production Depreciation

The *units of production* depreciation expense equals the cost less salvage value divided by the usage rate. This method also writes off the entire cost.

Declining Balance Depreciation

The *declining balance* method usually uses double the straight line rate in the first year. It then applies this same percentage to the remaining book value in each of the other years of the asset's useful life. In this method no allowance is made for salvage value, and the method *always* leaves a balance. It is the most commonly used method of accelerated depreciation.

Sum-of-the-Years Digits Depreciation

The *Sum-of-the-Years Digits* (SYD) depreciation expense is found by using the SYD = 5+4+3+2+1 = 15 as the denominator in a fraction whose numerator is the highest year. These calculations are shown in Table 3-1. This method allows larger deductions in earlier years.

Year	Calculation	Expense
1	(5/15) ($100,000) =	$33,333.33
2	(4/15) ($100,000) =	$26,666.67
3	(3/15) ($100,000) =	$20,000.00
4	(2/15) ($100,000) =	$13,333.33
5	(1/15) ($100,000) =	$ 6,666.67

Table 3-1 Sum-of-the-Years Digits Depreciation

A firm can use any depreciation method (or even all of them, on different classes of assets) for its reports to shareholders. Before 1980 any of the above mentioned accounting depreciation methods could also be used for tax purposes as well as for accounting purposes.

For assets placed into service after 1980 and before 1987, the *Accelerated Cost Recovery System* (ACRS) was required for tax purposes. It was similar to the modified accelerated cost recovery system (MACRS) which replaced it.

The Modified Accelerated Cost Recovery System (MACRS)

Only one method, the *Modified Accelerated Cost Recovery System* (MACRS), is now allowed by law for calculating depreciation expense for tax purposes. Thus, it is the *only relevant method for cash flow purposes.*

The modified accelerated cost recovery system assigns tangible assets placed into service after 1986 to one of 8 classes. The tax deductible expense is equivalent to the use of the double declining balance method over 3, 5, 7 or 10 years and to the 150% declining balance method over 15 or 20 years. There are also classes for residential real property and non residential real property with a class life of 27.5 years or more. For all the MACRS classes, when the tax deductible expense calculated by the straight-line method exceeds the amount calculated by the declining balance method, MACRS requires a switch to the straight-line method. Thus, the first cost of the asset is fully depreciated.

Assets are assumed to be placed in service in the middle of the first year thus reducing the deduction in the first year and making the deduction in the second year larger than would otherwise be the case.

For intangible property, the straight-line method is used. For more information consult IRS Publication 534.

Net Cash Flow

A firm's cash flows are generally equal to cash from sales, minus cash operating costs, and minus taxes. Depreciation is an operating cost but it is not a cash expenditure; it is merely an accounting entry. Thus, a firm's net cash flow in any accounting period can be approximated by adding its depreciation to its earnings after taxes. (There may also be some other adjustments needed, but this adding back of depreciation is the commonly used method for estimating cash flow).

It is usually assumed that the firm will continue with current operations, then add the new project. Hence, the relevant cash flows of the project are usually identified by comparing the situation without the project with the expected situation if the project is implemented.

In project analysis the timing of the cash flows is important, but we must compromise between accuracy and simplicity. Therefore, it is customary to assume that all project cash flows occur at the end of each year and forecasts are normally made for the several years of the economic life of the project. One method of forecasting is regression analysis, which is described in Appendix 3A.

Inflation

Suppose that costs are escalating at a given annual inflation rate and you would like to know how much these costs will be in the 25th year. We use an ESCALATE worksheet to demonstrate compound interest and compounding with spreadsheet software. This worksheet is on the diskette which accompanies this book. It is shown in Figure 3-1. To construct this worksheet, you need only type in the entries for the first 13 rows, including formatting the cells, if desired. Column A can be completed with the **Data Fill** command. Column B can be completed with the **Copy** command. Figure 3-2 shows the key formulae which produced Figure 3-1. Although the ESCALATE worksheet was designed primarily for estimating future costs, it could also be used to estimate future revenues if these were growing at a constant rate.

	A	B
1	ESCALATE	B36
3	INPUT DATA	
4		= Current cost
5		= Inflation rate
7	OUTPUT DATA	
8	1	= One + inflation rate
10	Period	Inflation adjusted cost
11	0	$0.00
12	1	$0.00
13	2	$0.00
14	3	$0.00
36	25	$0.00

Figure 3-1 An Escalate Worksheet

	A	B
1	ESCALATE	B36
4		= Current cost
5		= Inflation rate
8	1+$A$5	= One + inflation rate
10	Period	Inflation adjusted cost
11	0	+$A$4
12	1	+$A$4*$A$8
13	2	+$A$4*$A$8^$A13
14	3	+$A$4*$A$8^$A14
36	25	+$A$4*$A$8^$A36

Figure 3-2 Formulae for Escalate Worksheet

Appendix 3A

REGRESSION ANALYSIS

Most spreadsheets have built-in regression features as part of their data base management capabilities. Regression analysis fits a regression line to an existing data base consisting of a dependent or Y variable and an independent or X variable. This regression line, with intercept **a** and slope **b**, is estimated by minimizing the sum of squared errors Σe_i^2 (least squared errors).

An error e_i is the difference between an observed value Y_i of the dependent variable and the corresponding value $\hat{Y}_i$ "predicted" by the regression line.

The predicted value $\hat{Y}_i$ of the dependent variable is obtained by calculating

$$\hat{Y}_i = a + bX_i,$$

where X_i is the observed value of the independent variable associated with the observed value of the dependent variable.

Multiple regression analysis extends the concept of least squared errors to fit a regression surface to an existing data base consisting of a dependent variable and more than one independent variable.

The REGRESS worksheet discussed in this section is designed to handle up to 16 independent variables and 400 observations. However, the example presented later uses only one independent variable to make a forecast of sales revenues. (Regression analysis is, of course, just one of several techniques that can be used to forecast the value of a variable.)

Now load your spreadsheet software and retrieve the REGRESS.WKS file from your working copy of the template diskette that accompanies this book. The REGRESS worksheet that appears on your monitor screen should be similar to Figure 3A-1. Column width for each column has been preset globally at 14, instead of the standard 9, using the ⎡/⎤⎡W⎤⎡G⎤⎡C⎤⎡1⎤⎡4⎤⎡Enter⎤ command.

	A	B	C	D	...	R
1	REGRESS			R16	...	
2	Regression				...	
15	Dependent		Independent	Independent	...	Independent
16	Variable		Variable #1	Variable #2	...	Variable #16

Figure 3A-1 REGRESS Worksheet

Data Input

The REGRESS worksheet is preconfigured to expect data for the dependent variable in column A beginning with cell A17 and data for the first independent variable in column C beginning with cell C17. If more than one independent variable is used, data should start in cell D17 for independent variable #2 through R17 for independent variable #16.

Although rows 3-14 are omitted from Figure 3A-1, only row 14 is truly blank. If you write anything in the range of cells A4 through R13, it will probably cause problems with obtaining a proper display of a regression output.

Note that column B is left blank. In our own work, we have found that we frequently need to perform a transformation, such as the natural logarithm, on the dependent variable before running a regression. Such a transformation can be done in column B, and the transformed data can then be used as the dependent variable in the regression.

(Alternatively, we have found it very useful to calculate the "predicted" values $\hat{Y}_i$ associated with the observed values of the independent variables.)

Regression Output

The REGRESS worksheet is further preconfigured to display regression output in the range A4..R13, as well as to compute the Y intercept or constant for the regression line.

The formulae for the REGRESS worksheet are given in Figures 3A-2A through 2C. *Note that these formulae are all in the regression output range A4..R13.* We carefully picked empty cells in this range to augment the somewhat limited statistics that Lotus 1-2-3 release 2.2 provides.

In particular, we wanted to calculate the t values associated with the coefficients of the independent variables and the F ratio associated with the overall regression equation. *If you are using a different spreadsheet, you may need to move and/or revise these formulae.*

The formulae in cells A10 and A13, respectively, of Figure 3-2A display the labels F Ratio and t Value(s) if any one of the cells A4 through A9 is *not* empty. Otherwise, the screen displays for cells A10 and A13 are blank.

	A
10	@IF(@COUNT($A$4..$A$9)=0," ", "F Ratio")
13	@IF(@COUNT($A$4..$A$9)=0," ", "t Value(s)")

Figure 3A-2A Formulae for the REGRESS Worksheet

The formula in Figure 3A-2B displays the t value for the coefficient of the first independent variable if either of the cells C11 and C12 is *not* empty. Otherwise, the display for cell C13 is blank. The formulae for cells D13 through R13, which are not shown, are identical to the formula in cell C13, except that C is replaced by D through R respectively.

	C
13	@IF(@COUNT(C$11..C$12)=0," ", C$11/C$12)

Figure 3A-2B Formula for the REGRESS Worksheet

The formula in Figure 3A-2C displays the F ratio for the overall regression equation if any of the cells D4 through D9 is *not* empty. Otherwise, the video display for cell D10 is empty.

	D
10	@IF(@COUNT($D$4..$D$9)=0," ", ($D$7*$D$9)/((1-$D$7)*($D$8-$D$9-1)))

Figure 3A-2C Formula for the REGRESS Worksheet

	A	B	C
17	100		1
18	106		2
19	109		3
20	115		4
21	119		5
22	126		6
23	130		7

Figure 3A-3 REGRESS Worksheet
Revenues versus Time

Using the REGRESS Worksheet

To illustrate the use of the REGRESS worksheet, type in the data shown in Figure 3A-3. Sales revenues (thousands omitted) are shown in cells A17 through A23. Years are shown in cells C17 through C23.

Now type ⒡⑤ⓐ①ⒺⓃⓉⒺⓇ to go to cell A1. (⒡⑤ is the Go To key.)

Rename the worksheet ESTSALES (for estimated sales) by typing ⒺⓈⓉⓈⒶⓁⒺⓈⒺⓃⓉⒺⓇ .

Use the right arrow ⟶ key three times to scroll to cell D1. Now type Ⓓ②③ⒺⓃⓉⒺⓇ . This indicates that the print range for this worksheet is A1..D23.

To run the regression of sales revenues on time, invoke the **Data Regression** command by typing ⑦ⒹⓇⒺⓃⓉⒺⓇ .

The top part of the display on your video screen should now look like Figure 3A-4 with **X-Range** highlighted on line 2 of the control panel.

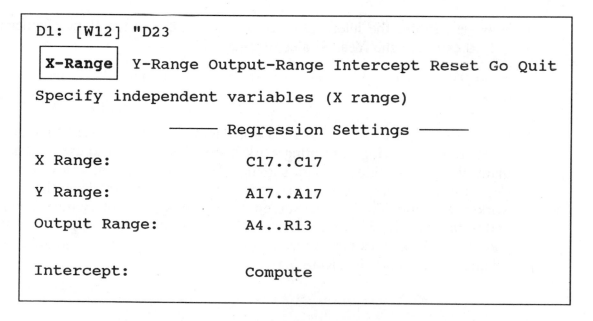

```
D1:  [W12]  "D23

 X-Range   Y-Range Output-Range Intercept Reset Go Quit

Specify independent variables (X range)

                 ——— Regression Settings ———

X Range:              C17..C17

Y Range:              A17..A17

Output Range:         A4..R13


Intercept:            Compute
```

Figure 3A-4 Regression Settings

If **X-Range** is highlighted, type (Enter) . Alternatively, whether X-Range is highlighted or not, type (X) . The cursor should now be on cell C17. If it is not, scroll to cell C17. (The Go To (F5) key is disabled when setting ranges.) Now type (.) and use the down arrow (↓) key to scroll to cell C23. (If there were four independent variables with ten observations, for example, you would scroll to cell F26. Column F means that you have four independent variables and row 26 less 16 rows of non data means that you have ten observations.) Press(Enter) to complete the setting of the **X-Range** as C17..C23.

Setting the **Y-Range** is similar to setting the **X-Range**. First, press (Y) . The cursor should be on cell A17. Now type (.) and use the down arrow (↓) key to scroll to cell A23. Finally, press (Enter) to complete the setting of the **Y-Range** as A17..A23.

The **Output-Range** should already appear as A4..R13 as shown in Figure 3A-4. If it is different from this, it can be adjusted in a manner analogous to the **X-Range** and **Y-Range** adjustments you just made.

Intercept should appear as **Compute** in Figure 3A-4. The only other possibility is **Zero**. If **Zero** appears, type (I)(C) and **Compute** will appear.

You are now ready to run the regression. Type (G) . The regression output should appear on your monitor as shown in Figure 3A-5.

Now we can use the intercept or constant **a** = 95 and the slope coefficient **b** = 5 to estimate the Year 8 sales revenues, Y_8:

$\hat{Y}_8 = 95 + 5(8) = 135$. Since thousands were omitted when sales revenues were entered, 135 represents estimated sales revenues of \$135,000 in Year 8.

If you wish to save the ESTSALES worksheet or to print it, see Chapter 1 for instructions on saving or printing worksheets. If you intend to use a worksheet many times with different data sets in one working session, you will need to clear the worksheet with data from the screen and retrieve the original worksheet again. To clear the screen, type (/)(W)(E)(Y) . Alternatively, you can add to the REGRESS worksheet the \R macro discussed in Appendix C at the end of the book. This macro resets any REGRESS worksheet with data to the original state shown in Figure 3A-1.

	A	B	C	D
4		Regression	Output	
5	Constant			95
6	Std Err of Y Est			0.894427191
7	R Squared			0.9943181818
8	No. of Observations			7
9	Degrees of Freedom			5
10	F Ratio			875
11	X Coefficient(s)		5	
12	Std Err of Coef.		0.1690308509	
13	t Value(s)		29.580398915	

Figure 3A-5 Regression Output

Tests of Significance

Suppose you want to test whether the slope coefficient **b** is significantly different from zero. The calculated t value for our example is 29.58039891 (displayed in cell C13 of Figure 3A-5). The degrees of freedom associated with this t calculation are five (displayed in cell D9). The table t value (from virtually any standard statistics textbook) for a two-tailed t test at the one percent level of significance (0.5% in each tail) with five degrees of freedom is 4.032. Since the magnitude or absolute value of the calculated t exceeds the table t, we conclude that the slope coefficient is significantly different from zero. That is, we reject the null hypothesis that **b** equals zero and accept the alternate hypothesis that **b** is different from zero.

The F ratio allows us to test the overall regression equation. In the F test, the null hypothesis is that the coefficients of all the independent variables are zero. The alternative hypothesis is that at least one of the coefficients is different from zero. Clearly, when there is only one independent variable, the F test should give results that are consistent with a two-tailed t test at the same level of significance.

We will use our example to illustrate a relationship between t and F that always holds in the one independent variable case and guarantees the consistency between the t and F tests in this case. The calculated F ratio is 875 (displayed in cell D10 of Figure 3A-5). The F ratio has degrees of freedom in both the numerator and denominator. The degrees of freedom for the numerator equals the number of independent variables, which is one in our example. The denominator of the F ratio has the same degrees of freedom as the t value (displayed in cell D9 of Figure 3A-5), which is five in our example.

The table F ratio (from virtually any standard statistics textbook) for one and five degrees of freedom at the one percent level of significance is 16.26. Since the calculated F of 875 exceeds the table F of 16.26, we reject the null hypothesis and accept the alternate hypothesis, just as we did for the t test.

Consistency between the F test and the t test is guaranteed in the one independent variable case because calculated F always equals calculated t^2 and table F always equals table t^2.

In our example, calculated $F = 875 = (29.58039891)^2 = $ calculated t^2 and table $F = 16.26 = (4.032)^2 = $ table t^2.

Thus, when there is only one independent variable, the F ratio provides no additional information than what is provided by the t value for the slope coefficient.

When there is more than one independent variable, however, the F test and the t tests may give apparently conflicting results. It is possible for the t tests to indicate that the coefficients of all of the independent variables are *not* significantly different from zero and for the F test to indicate that at least one of the coefficients (the F test does not tell us which one) is significantly different from zero.

That is, the t tests tell us that individually the coefficients are insignificant, whereas, the F test tells that the regression equation is significant overall.

It is the F test that must be used to test the overall significance of the regression equation, not the individual t tests, because the F test takes account of all the covariances among the estimated regression coefficients.

Chapter 4

DISCOUNTED CASH FLOW ANALYSIS

This chapter introduces discounted cash flow calculations. Topics covered are weighted average cost of capital, minimum acceptable rate of return, discounted cash flow analysis, internal rate of return, present worth comparisons, equivalent annual worth comparisons, and multiple internal rates of return.

Key Concepts

The *weighted average cost of capital* (WACC) is a composite rate which represents the after-tax cost of all acquired funding for a firm. Capital is usually obtained from several sources, each of which has a cost. The composite cost is weighted according to the market proportion of funds from each source. While the WACC can be calculated based on historical information, such information is *not* relevant for investment decision-making purposes, since the relevant cost is the *opportunity cost*, or potential return foregone by investing in the project. This opportunity cost is the return which could have been obtained by investing elsewhere. Hence, market rates measure the opportunity cost of funds used in any investment project. Since we could have obtained this alternative rate of return but chose instead to invest in the project, we want to obtain a rate of return at least this high from the project.

To summarize, the firm's WACC is the sum of the market or opportunity costs of all funds that the firm uses (*not just funds raised currently*), weighted by their market, *not book*, proportions.

The *minimum acceptable (or attractive) rate of return* (MARR) is the lower limit of the minimum rate of return that investors must expect to earn from an investment project for them to be willing to finance it. This is usually at least the WACC, but it should be higher if the project is of greater than average risk. Similarly, a project's opportunity cost of capital, and hence, its MARR could be less than the firm's WACC if the project were below average risk.

Discounted cash flow analysis is the method used to evaluate investment projects which require outlays of capital (at one or more points in time) to generate future cash flows at other points in time. Cash flows at different times have different present worths (values) because of the time value of money. In particular, a dollar today can be invested to earn interest. Hence, it is more valuable than a dollar to be received some time in the future. Thus, future cash flows must be *discounted* back to the present so that the various cash flows can be compared on an equivalent basis.

The *internal rate of return* (IRR) is a widely used method for evaluating investment projects. The IRR is calculated by solving for the discount rate that equates the sum of the present worths of the cash flows to zero. This method works well when there is only one cash outflow, followed by cash inflows. However, in projects that require another cash outflow after the cash inflows have started, such as overhauling a generator, the change in sign of the cash flow stream causes the equation to potentially have two positive roots. Since there is no way to distinguish between these roots, there is a multiple rate of return problem.

The *present worth* (or present value) method overcomes the multiple rate of return problem since it can handle any number of cash outflows and inflows in any order and will always give one and only one solution.

The *net* present worth is calculated by discounting the cash inflows at the WACC or MARR and then subtracting the first cost (initial outlay) or discounted value of all the cash outflows. If the result is positive, the investment should be made.

Equivalent annual worth comparisons produce results comparable with present worth and are especially useful in comparing investments projects which *do not* have the same useful lives.

WACC and MARR

Figure 4-1 presents a very simple worksheet for calculating the weighted average cost of capital, WACC. Note that the label D14 in cell D1 in Figure 4-1 is simply a reminder that the print range is from A1 to D14. Figures 4-2A and 4-2B contain the formulae for the WACC worksheet. Figure 4-3 gives an example of the calculation of the WACC.

To use the WACC worksheet, first load your spreadsheet software and then retrieve the file called WACC.WKS from your working copy of the diskette that accompanies this book.

	A	B	C	D
1	WACC			D14
2	WEIGHTED AVERAGE COST OF CAPITAL			
3				
4		Tax Rate		
5		(for example, enter 40% as .4		
6				
7		Percentage	Percentage	Contribution
8	Component	Weight	Cost	to WACC
9				
10	Debt (Before Tax)			
11	Preferred Stock			
12	Common Equity			
13				
14	Sum of Weights =	WEIGHT ERROR	WACC =	

Figure 4-1 WACC Worksheet

Data Entry and Error Trapping

Data for the firm's tax rate is entered in cell A4 in either percentage or decimal form. For example, 40 percent would be entered as either **40%** or **.4** and would be displayed as **40.00%** as shown in Figure 4-3.

Data for the weight or fraction of financing that comes from each component type is also entered in either percentage or decimal form. The debt, preferred stock, and common equity weights are entered in cells B10 through B12, respectively. These component weights must sum to 100% or one. (Even though percentages are displayed in cells B10 through B12, internally they are stored as their decimal equivalents.) Note that the weights shown in Figure 4-3 do satisfy this requirement.

If the weights do not sum to one, the worksheet will display WEIGHT ERROR as shown in cell B14 of Figure 4-1. The formula that performs this *error trapping* is shown in Figure 4-2A.

Ideally, this formula could be simply written as: @IF(@SUM($B$10 ..$B$12)=1.0,@SUM($B$10..$B$12), "WEIGHT ERROR"). This logical @IF statement says that if the sum of the component weights is 1.0, display 100% in cell B14. (100.00%, rather than 1.0 is displayed because cell B14 is formatted in percentage with two decimal places.) Otherwise, display WEIGHT ERROR.

Unfortunately, when you try to type this statement into some spreadsheets, the software changes the floating point 1.0 to the integer 1. Then when the sum of the weights equals the floating point 1.0, the software apparently says that this is not the integer 1 and displays the error message WEIGHT ERROR.

Our solution to this software peculiarity was to employ the logical #AND# function. If the sum of the weights is less than 1.00001 and greater than 0.99999, then display the sum. Otherwise, display WEIGHT ERROR.

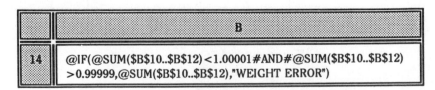

Figure 4-2A Formula for WACC Worksheet

	D
10	@IF(@COUNT($B10..$C10)=0," ",@ROUND((1-$A$4)*$B10*$C10,4))
11	@IF(@COUNT($B11..$C11)=0," ",@ROUND($B11*$C11,4))
12	@IF(@COUNT($B12..$C12)=0," ",@ROUND($B12*$C12,4))
14	@IF(@COUNT($B$10..$B$12)=0," ",@SUM($D$10..$D$12))

Figure 4-2B Formula for WACC Worksheet

	A	B	C	D
1	WACCEXAM			D14
4	40.00%	Tax Rate		
7		Percentage	Percentage	Contribution
8	Component	Weight	Cost	to WACC
10	Debt (Before Tax)	30.00%	10.00%	1.80%
11	Preferred Stock	10.00%	12.00%	1.20%
12	Common Equity	60.00%	14.20%	8.52%
14	Sum of Weights =	100.00%	WACC =	11.52%

Figure 4-3 WACC Worksheet Example

Error trapping should be widely employed in worksheets used in working environments. However, rather than clutter up this book with numerous examples of error trapping, we have chosen to illustrate this important engineering principle in just this one simple worksheet.

Data for the component financing costs of debt, preferred stock, and common equity are entered as either a percentage or decimal in cells C10 through C12, respectively, and are displayed as percentages as shown in Figure 4-3. For example, if 10 percent is the before tax cost of debt, then 10% or .1 should be entered in cell B10 and 10.00% will be displayed in cell C10.

The contribution of each component of financing to the WACC is calculated in cells D10 through D12. The WACC is the sum of these contributions and is displayed as 11.52% in cell D14 of Figure 4-3 for our numerical example.

The formulae for cells D10 through D12 are shown in Figure 4-2B. Except for row indexing and the tax factor **(1-$A$4)**, these three formulae are identical. The tax factor appears in the debt contribution because interest is tax deductible. For example, $6 in earnings after taxes would support an increase in interest of $10, if the firm is in the 40 percent tax bracket. That is, a $10 increase in interest reduces taxes by $4 and reduces earnings after taxes by $6. Hence, a before tax cost of debt of 10 percent results in a 10%(1-.4) = 6% after tax cost of debt for a firm in the 40 percent tax bracket.

The @ROUND function is used in the cells D10 through D12. Recall that the reason for using the @ROUND function was discussed in the AMORTGEN worksheet in Chapter 2. Note that four decimal places are needed in the @ROUND function because a display of **XX.XX%** is stored internally as **.XXXX** If only two places were used, 2.00%, 1.00%, and 9.00% would be displayed in cells D10 through D12, respectively, and the WACC displayed in cell D14 would be 12.00%.

The @IF and @COUNT functions are also used in cells D10 through D12 and in D14. These functions were used in the AMORTGEN worksheet in Chapter 2 and the REGRESS worksheet in Appendix 3A, respectively. These functions make the displays in cells D10 through D12 and D14 blank before any data are entered in cells B10 through C12.

The WACC worksheet presumes that the user already has obtained the necessary input data. A more sophisticated WACC worksheet could incorporate formulae for calculating the cost of common equity, preferred stock, and debt.

Estimating Component Costs

One way of *estimating the cost of common equity from retained earnings* (profits which are not paid out in dividends to stockholders) is the constant growth rate method. For many public utilities, this method works well and is frequently used in regulatory rate hearings. In this method, the cost of common equity k_c is comprised of (1) a current dividend yield (D_1/P_0) and (2) the constant growth rate g of dividends per share. $k_c = (D_1/P_0) + g$

Note that D_1 represents dividends per share for the current year and P_0 represents the stock price at the start of the current year. If common equity is raised by issuing (floating) new stock rather than from additions to retained earnings, the price P_0 in the constant growth formula must be multiplied by (1-F), where F is an after-tax flotation cost percentage for the new stock.

In the constant growth model, the dividends per share D_t for period t are simply $(1+g)$ times the dividends per share D_{t-1} for period $t-1$. Thus, all dividends per share for any period t are related to the dividends in an initial period 0 by

$$D_t = D_0(1+g)^t \text{ or } \ln D_t = \ln D_0 + [\ln(1+g)]t$$

To estimate g, first regress the $\ln D_t$ against t using the REGRESS worksheet from Appendix 3A. The estimated slope b of the regression line $\ln D_t = a + bt$ is an estimate of $\ln(1+g)$. Thus, an estimate of g is $(e^b - 1)$.

Preferred stock usually pays a constant level of dividends. Hence, the constant growth rate formula simplifies to the zero growth case when it is applied to preferred stock. The estimated cost k_p of preferred stock is comprised only of a dividend yield (D_p/P_p). Of course, many corporations do not have preferred stock. For such firms, one simply enters 0 for the preferred stock weight and the percentage (%) preferred stock cost in cells B11 and C11, respectively.

The *before tax cost of debt* is essentially the *yield to maturity* (YTM) on a firm's outstanding bonds. If the bonds are actively traded, the YTM is published in most financial newspapers on a regular (at least weekly) basis.

It is generally agreed that the minimum acceptable (or attractive) rate of return, MARR, is a lower limit for investment acceptability. Many organizations deliberately choose to set the MARR above the WACC. The MARR, however, could be set lower than the firm's WACC for a particular project that is less risky than the firm's average project. Nevertheless, the WACC worksheet is useful in establishing the MARR.

DCFA and IRR

Figure 4-4 presents the discounted cash flow analysis, DCFA, worksheet, which you can retrieve as the file DCFA.WKS from the accompanying diskette.

Similar to the loan amortization worksheet that appeared in Figure 2-5 of Chapter 2, the DCFA worksheet automatically adjusts the number of rows that appear in the worksheet. *The user entry in cell B9 controls the number of rows.* This is the number of periods, N, for which there are cash flows. The minimum number of periods is 1 and the maximum is 25. We limited the maximum number of periods to 25 because it is unusual for an engineering economics problem to require more than 25 periods. Furthermore, the larger the maximum number of periods, the slower the model will run, no matter how fast your PC is.

	B	C	D	E	F	G
1	DCFA					G45
4	RESULTS:	EAW		IRR		NPW
5		ERR		ERR		$0.00
7	DATA ENTRY: Cells B9, B12, & B15 plus cash flows starting in C20.					
9		= N, the Number of Periods				
10		(Maximum = 25, Minimum = 1)				
12		= i, the Periodic Interest Rate				
13		(for example, enter 10% or .1)				
15		= Estimate of IRR, the Internal Rate of Return				
16		(Estimate must be between 0% and 100% or 0 and 1)				
18	Period	Cash Flow	x	(P/F,i,N)	=	PW of CF
20	0			1.000000		$0.00

Figure 4-4 DCFA Worksheet

The user enters a periodic interest rate, i, in cell B12 of the DCFA worksheet. Normally, the MARR would be used. Note that the periodic interest rate is entered in either percentage or decimal form. For example, 10 percent would be entered as either 10% or .1 and displayed as 10.00%.

An estimate of the internal rate of return, IRR, must be entered in cell B15. The entry must be no larger than 1 or 100% and no smaller than 0. From experience, .5 seems to work better than 0 or 1. Of course, if you have a better estimate, you should use it.

Two cautionary notes about the IRR must be made: (1) Some projects may not have a positive IRR and (2) some projects may have multiple positive IRRs. (Later in this chapter an example of a project with multiple IRRs is given in Figures 4-8A and B.)

In most spreadsheets, problems without a positive IRR lead to the built-in error message, **ERR**, appearing in cell E5, instead of a calculated value of the IRR. *(This initially may cause some confusion among engineers, as the symbol* **ERR** *is generally used in engineering economics to mean the external rate of return.)* Since there is no data entered in the DCFA worksheet in Figure 4-4, this also causes the error message, **ERR**, to appear in cell E5. (Some spreadsheets, however, do use **ERROR** for the error message.)

Additional data entry consists of the cash flows. These are entered in cells C20 to C45, starting with the first cost in cell C20 and proceeding down the column C until the last cash flow is entered. Note that the first cost is entered with a negative sign because it is a cash outflow. For example a first cost of $100 must be entered as **-100** in cell C20 and will be displayed as **($100.00)**, where the parentheses indicate a negative number. *(Do **not** enter the first cost, however, as (100) or ($100) as this causes problems with some spreadsheet software.)*

Note that the DCFA worksheet in Figure 4-4 starts with column B. Column A is hidden from the user. This is done with the hide command, /WCH . In response to the query for a range, A1 is entered. If you want to see what is in column A, you must invoke the display command, /WCD . Then A1 is entered as the range and the cursor is moved to column A with the left arrow key, ← .

Figures 4-5A and 4-5B display the formulae for the DCFA worksheet, including those in column A. Rows 1 to 4 and rows 6 to 18 are not shown in Figures 4-5A and 4-5B because they contain only textual material which was already displayed in Figure 4-4. Row 19 is not shown as cells B19 through G19 contain only the shorthand expression \- which fills the cells with a dashed line. Rows 22 to 44 are not shown, as the formulae are generated by copying row 21. When the formulae are copied, some parts of the formulae are automatically indexed (increased by 1 per row) by most spreadsheets. For example, $A21 becomes $A22 in row 22, $A23 in row 23, and so forth. Row 45 is also generated by copying row 21. Comparison of row 45 to row 21 in Figures 4-5A and 4-5B illustrates precisely where indexing is occurring.

	A	B	C
5			($G$5*$B$12)/(1-1/(1+$B$12) ^ $B$9)
20	0	@IF($B$9>=$A20,$A20,"")	
21	1+$A20	@IF($B$9>=$A21,$A21,"")	
45	1+$A44	@IF($B$9>=$A45,$A45,"")	

Figure 4-5A Formulae for DCFA Worksheet

	E	G
5	@IRR($B$15,$C$20..$C$45)	@SUM($G$20..$G$45)
20	@IF($B$9>=$A20,1/(1+$B$12) ^ $A20,"")	@IF($B$9>=$A20,$C20*$E20,"")
21	@IF($B$9>=$A21,1/(1+$B$12) ^ $A21,"")	@IF($B$9>=$A21,$C21*$E21,"")
45	@IF($B$9>=$A45,1/(1+$B$12) ^ $A45,"")	@IF($B$9>=$A45,$C45*$E45,"")

Figure 4-5B Formulae for DCFA Worksheet

Starting with column A, note that cell A20 contains the number **0**. This means that the formulae in cells A21 through A45 generate the numbers 1 through 25, respectively. However, we do not want to see all of these numbers if we have only a 2 period problem. This difficulty is solved by hiding column A and using the @IF function in cells B20 to B45 of column B.

In particular, note the formula **@IF($B$9>=$A20,$A20,"")** in cell B20. Recall that the number of periods, N, for a problem is entered in cell B9. (Since we have yet to enter any value in cell B9, most spreadsheets interpret this as a value of 0 in cell B9.) If the value of N exceeds or equals the current period contained in cell A20, then the contents of A20 are displayed in cell B20. Otherwise, this cell will display whatever is contained between the quote marks, " ".

Since there is nothing between the quote marks, the display will be blank if N is less than the current period. Thus, for cell B20, **0** should always be displayed because N effectively is 0 before data is entered and ≥ 1 in any practical problem. Cells below B20 will display successive periods until period N is reached.

Moving to column C, note that the formula **($G$5*$B$12)/(1-1/ (1+$B$12)^$B$9)** in cell C5 calculates the EAW of a project.

Like cell B19, cell C19 has the expression **\-**, which fills this cell with a dashed line.

As mentioned earlier, cells C20 to C45 are for the entry of the cash flows. Thus, no formulae appear there. Column D is not shown in Figure 4-5B because it is empty except for **\-** in cell D19.

In column E, cell E5 contains the function **@IRR ($B$15,$C$20.. $C$45)**, which calculates the IRR based on the initial estimate in cell B15 and the cash flows in cells C20 through C45. Cells E20 to E45 contain @IF statements.

For example, cell E20 contains the function **@IF($B$9>=$A20,1/ (1+$B$12)^$A20,"")**. This function returns the value of the PW factor for period 0, if period 0 is less than or equal to the maximum number of periods N in Cell B9. Otherwise, the display is blank for this cell.

Again, since N effectively is 0 before data is entered and $\geq$ 1 in any practical problem, the PW factor of 1.000000 should always be displayed in cell E20.

The cells E21 to E45 display the PW factors for successive periods until period N is reached.

Column F is omitted from Figure 4-5B because it is identical to column D.

Finally, in column G the formula **@SUM($G$20..$G$45)** in cell G5 calculates the NPW of the project by summing the PWs for periods 0 to 25, which are calculated by the formulae in cells G20 to G45.

NPW and EAW Comparisons

When two competing projects have the same economic life, the net present worths of the two projects can be directly compared. The project with the higher NPW is normally chosen.

Suppose, however, the economic lives differ and the projects can be repeated. Then selecting the project with the higher NPW may lead to the wrong decision. In this situation a common study period should be selected. The NPWs can be calculated for this common study period, and the project with the higher common study period NPW is chosen. Alternatively, one can compare equivalent annual worths, provided each time a project is repeated its cash flows are identical.

Figures 4-6 and 4-7 show the EAWs, IRRs, and NPWs for two competing projects.

Figure 4-6 contains a three-period project with cash flows of -$100, $60, $40, and $40 in periods 0 through 3, respectively. Its IRR is 20.64% and its NPW is $17.66 at a 10% MARR.

Figure 4-8 contains a four-period project with cash flows of -$100, $50, $40, $40, and $18 in periods 0 through 4. Note that it has a higher IRR, 20.77%, and a higher NPW, $20.86, than the three-period project. (Also note that the displayed NPW of $20.86 is one cent greater than the sum of the displayed cash flows in cells G20 through G24. This rounding problem was discussed in Chapter 2 in the Loan Amortization section. This problem can be solved by replacing **$C20*$E20** in cell G20 with **@ROUND($C20*$E20,2)** and making the equivalent replacements in cells G21 through G45.)

Hence, if the projects were not repeatable, the four-period project should be selected because it has higher NPW.

However, if the projects are repeatable, the three-period project is preferable to the four-period project. This is because the EAW of the three-period project is $7.10, which exceeds the EAW of $6.58 for the four-period project. This means that if the projects were repeated in perpetuity, the NPW over an infinite horizon is $7.10/.10 = $71 for the three-period project versus only $6.58/.10 = $65.80 for the four-year project.

Alternatively, if one examines the two projects over a common twelve-year study period, the study-period NPW is $7.10(P/A,10,12) = $48.38 for the three-period project versus $6.58(P/A,10,12) = $44.83 for the four-year project.

To construct the worksheet example shown in Figure 4-6, simply enter in the DCFA worksheet the data displayed in cells B9, B12, B15, and C20 to C23.

Note that if you wish to save this DCFA worksheet example, you should give it a new name. Otherwise, you will write over the existing DCFA worksheet. For this reason, we changed the cell A1 from **DCFA** to **DCFA3PD** to suggest a name you may want to use. We also changed the label from G45 to G23 in cell G1 to indicate that the print range is now B1 to G23. (Remember, in the DCFA worksheet, column A is hidden.)

	B	C	D	E	F	G
1	DCFA3PD					G23
4	RESULTS:	EAW		IRR		NPW
5		$7.10		20.64%		$17.66
9	3	= N, the Number of Periods				
12	10.00%	= i, the Periodic Interest Rate				
15	50.00%	= Estimate of IRR, the Internal Rate of Return				
18	Period	Cash Flow	x	(P/F,i,N)	=	PW of CF
20	0	($100.00)		1.000000		($100.00)
21	1	$60.00		0.909091		$54.55
22	2	$40.00		0.826446		$33.06
23	3	$40.00		0.751315		$30.05

Figure 4-6 DCFA Three-period Worksheet

	B	C	D	E	F	G
1	DCFA4PD					G24
4	RESULTS:	EAW		IRR		NPW
5		$6.58		20.77%		$20.86
9	4	= N, the Number of Periods				
12	10.00%	= i, the Periodic Interest Rate				
15	50.00%	= Estimate of IRR, the Internal Rate of Return				
18	Period	Cash Flow	x	(P/F,i,N)	=	PW of CF
20	0	($100.00)		1.000000		($100.00)
21	1	$50.00		0.909091		$45.45
22	2	$40.00		0.826446		$33.06
23	3	$40.00		0.751315		$30.05
24	4	$18.00		0.683013		$12.29

Figure 4-7 DCFA Four-period Worksheet

Most spreadsheets have a built-in @NPV function. The @NPV function calculates the PW of cash flows for periods 1 through 25 directly from the undiscounted cash flows in cells C21 through C45 and the interest rate in cell B12. The first cost must be subtracted from this function to get the NPW.

Thus, **@NPV($B$12,$C$21..$C$45)+$C$20** should give the same NPW as the result displayed in cell G5. (Remember, cell C20 contains minus the first cost. So, that is why one must add the contents of that cell to the @NPV function.)

To test the @NPV formula, you will need a cell in which to enter the formula. Try cell H5.

Recall that in Chapter 1 we mentioned that most worksheets on the diskette are globally protected. In such worksheets, only the data entry cells, the cell (cell B1 in this case) which contains the name of a worksheet, and the cell which contains the print range (cell G1 here) are unprotected. To unprotect cell H5, make sure the cursor is on cell H5 and then type ⁄ R U ⏎ . Now type the @NPV formula in cell H5 and compare the results with cell G5.

Multiple IRRs

Figures 4-8A and 4-8B contain the results for a two-period problem with two positive internal rates of return. The cash flows are -$100, $330, and -$242 for periods 0, 1, and 2, which are entered in cells C20 through C22, respectively. The number of periods, N, is 2, which is entered in cell B9.

The periodic interest rate, i, is 46⅔%, which is entered as either 46.6667% or .466667 in cell B12 and is displayed as 46.67%. (This rate was selected solely for illustrative purposes, as it is the rate that maximizes the net present worth of this example.)

Note that the only data entry difference for the worksheets in Figures 4-8A and 4-8B is the estimated IRR in cell B15. In Figure 4-8A the initial estimate for the IRR in cell B15 is .5 or 50%, which leads to a calculated IRR of 10.00% in cell E5. In Figure 4-8B the initial estimate for the IRR is .054 or 5.4%, which leads to a calculated IRR of 120.00%. Depending on which spreadsheet software you use, you may need different initial estimates of IRR to generate these two final values.

This problem certainly illustrates the strange behavior of the IRR itself and the algorithm built into the spreadsheet used by the authors. Note that the lower initial estimate of IRR generated the higher final IRR!

	B	C	D	E	F	G
1	DCFAIRRS					G22
4	RESULTS:	EAW		IRR		NPW
5		$10.90		10.00%		$12.50
9	2	= N, the Number of Periods				
12	46.67%	= i, the Periodic Interest Rate				
15	50.00%	= Estimate of IRR, the Internal Rate of Return				
18	Period	Cash Flow	x	(P/F,i,N)	=	PW of CF
20	0	($100.00)		1.000000		($100.00)
21	1	$330.00		0.681818		$225.00
22	2	($242.00)		0.464876		($112.50)

Figure 4-8A DCFA Example, IRR = 10%

	B	C	D	E	F	G
1	DCFAIRRL					G22
4	RESULTS:	EAW		IRR		NPW
5		$10.90		120.00%		$12.50
9	2	= N, the Number of Periods				
12	46.67%	= i, the Periodic Interest Rate				
15	5.40%	= Estimate of IRR, the Internal Rate of Return				
18	Period	Cash Flow	x	(P/F,i,N)	=	PW of CF
20	0	($100.00)		1.000000		($100.00)
21	1	$330.00		0.681818		$225.00
22	2	($242.00)		0.464876		($112.50)

Figure 4-8B DCFA Example, IRR = 120%

Chapter 5

ADVANCED DCF TOPICS

This chapter discusses net present worth with taxes, sensitivity, risk, inflation, and replacement analyses (equal as well as unequal lives).

NPW with Taxes

Figure 5-1 presents a worksheet for calculating NPW with taxes, NPWTAX. To use the NPWTAX worksheet, first load your spreadsheet software and then retrieve the file NPWTAX.WKS from your working copy of the diskette that accompanies this book. Data for first cost (FC), additions to working capital (WC), estimated salvage value (S), useful economic life (N), modified accelerated cost recovery system (MACRS) class, tax rate, and the MARR are entered in cells A4 through A10, respectively. Note that the worksheet handles MACRS 3, 5, 7, 10, 15, and 20 year classes. Maximum life of an asset is limited to 25 years and must be greater than the selected MACRS class. Note also that rows 2, 3, 11, and 12 of the worksheet have been omitted from Figure 5-1. Rows 2 and 3 are empty except for the label "Net Present Worth with Taxes" in cell A2. Rows 11 and 12 have formulae related to the year of operation (see Figure 5-2A) but appear blank on the screen before data are entered into the worksheet.

	A	B	C	D
1	NPWTAX			Z27
4		First Cost, FC		
5		Additions to Working Capital, WC		
6		Estimated Salvage Value, S		
7		Useful Life, N (Warning: MACRS Class < N < 26)		
8		MACRS Class (Enter 3, 5, 7, 10, 15, or 20)		
9		Tax Rate (Enter, for example, 40% or .4)		
10		MARR (Enter, for example, 15% or .15)		
13	Revenues			
14	Operating Costs			
15	Operating Income			
16	Depreciation			
17	Taxable Income			
18	Taxes			
19	Net Income			
20	Operating NCF			
21	PW Factor			
22	PW NCF			
23	Total PW NCFs			
24	PW A-T Salvage			
25	PW Add. to WC			
26	NPW			
27	EAW			

Figure 5-1 NPWTAX Worksheet

In addition to the data entered in cells A4 through A10, data must also be entered for yearly revenues and operating costs. Yearly revenues must be entered in row 13, starting with column B for year 1. Column Z would be the ending column, if N takes on the maximum permitted value of 25 years. One way to estimate revenues is to use the REGRESS worksheet in Appendix 3A. Yearly operating costs must be entered in row 14, starting with column B. One way to estimate these costs is to use the ESCALATE worksheet in Chapter 3.

Column A of the worksheet contains no formulae. However, there is a hidden data entry in cell A11. Move the cursor to cell A11 and then press ⑦ⓇⒻⓉEnter . This will reveal that 0 is entered in cell A11. To again hide the data entry, make sure the cursor is still on cell A11 and press ⑦ⓇⒻⒽEnter . These commands perform essentially the same operations for an individual entry or range of entries as the display and hide commands perform for a column or row as discussed in Chapters 2 and 4 for the AMORTGEN and DCFA worksheets, respectively.

Figure 5-2A contains the formulae for column B of the worksheet. The first formula in Figure 5-2A is in cell B11. Recall that the value 0 is hidden in cell A11. Thus, the formula in cell B11 calculates the value 1. This value is used in computations but, like the 0 in cell A11, it does not appear on the screen because it is hidden.

The @IF function in cell A12 displays the label **YEAR 1** if the useful economic life, N, equals or exceeds 1 and displays an empty cell otherwise.

Cells B15 through B27 contain @IF statements that, like cell A12, display empty cells unless certain conditions are met. Since cell B16 contains the most interesting formula, it warrants further discussion.

First note that cell B16, which calculates depreciation, contains embedded @ROUND and @VLOOKUP functions. The @ROUND function is rounding the depreciation calculation to the nearest dollar.

If you wanted rounding to the nearest penny, then **...$A$4,0)** must be changed to **...$A$4,2)**. You would also have to change the display of column B to show currency in dollars and cents. This is accomplished by moving the cursor to cell B13 and pressing ⑦ⓇⒻⒸ②Enter . Then press ⨀ and move the cursor to cell B27 and press Enter . You also probably need to expand the width of columns B through Z from 12 to 15 by pressing ⑦ⓌⒼⒸ①⑤Enter .

	B
11	1+A$11
12	@IF($A$7>=B$11," YEAR 1"," ")
15	@IF($A$7>=B$11,B$13-B$14," ")
16	@IF($A$7>=B$11,@ROUND(@VLOOKUP($A$8,$A$28..$Z$33,B$11)*$A$4,0)," ")
17	@IF($A$7>=B$11,B$15-B$16," ")
18	@IF($A$7>=B$11,@ROUND($A$9*B$17,0)," ")
19	@IF($A$7>=B$11,B$17-B$18," ")
20	@IF($A$7>=B$11,B$16+B$19," ")
21	@IF($A$7>=B$11,1/(1+$A$10)^B$11," ")
22	@IF($A$7>=B$11,@ROUND(B$20*B$21,0)," ")
23	@IF($A$7>=$B$11,@SUM($B$22..$Z$22)," ")
24	@IF($A$7>=$B$11,@ROUND($A$6*(1-$A$9)/(1+$A$10)^$A$7,0)," ")
25	@IF($A$7>=$B$11,@ROUND($A$5*(1-1/(1+$A$10)^$A$7),0)," ")
26	@IF($A$7>=$B$11,-$A$4+$B$23+$B$24-$B$25," ")
27	@IF($A$7>=$B$11,($B$26*$A$10)/(1-1/(1+$A$10)^$A$7)," ")
28	0.33330
29	0.20000
30	0.14290
31	0.10000
32	0.05000
33	0.03750

Figure 5-2A Formulae for NPWTAX Worksheet

The @VLOOKUP function in cell B16 looks up the first year depreciation rate of 0.3333 in cell B28 if the MACRS class is 3 years, 0.2000 in cell B29 if the MACRS class is 5 years, and so forth. The MACRS class numbers 3, 5, 7, 10, 15, and 20 are hidden in cells A28 through A33, respectively. To see these hidden entries, scroll to the relevant cell and type `/` `R` `F` `T` `Enter` to reveal the contents of the cell.

Figure 5-2B illustrates the MACRS depreciation rates for the 3 year class in row 28 and the 5 year class in row 29. We did not print out all the rates for the other classes. However, they are in the worksheet, starting with the first year rate for the 7 year class in cell B30 and ending with the twenty-first year rate for the 20 year class in cell V33. Note that these MACRS rates are also hidden.

	B	C	D	E	F	G
28	0.33330	0.44450	0.14810	0.07410		
29	0.20000	0.32000	0.19200	0.11520	0.11520	0.05760

Figure 5-2B MACRS Depreciation Rates for NPWTAX Worksheet

Figure 5-2C contains the formulae for column Z of the worksheet. Compare the formulae in row 12 and in rows 15-22 in column Z with those in column B contained in Figure 5-2A. The main difference is that wherever B appears in the formulae in column B, it is replaced by Z in column Z. In row 11, but not in other rows, A is replaced by Y. A similar pattern holds for columns C-Y. For column Y, for example, Y would replace B and in row 11, X would replace A in the formulae.

	Z
11	1+Y$11
12	@IF($A$7>=Z$11," YEAR 25"," ")
15	@IF($A$7>=Z$11,Z$13-Z$14," ")
16	@IF($A$7>=Z$11,@ROUND(@VLOOKUP($A$8,$A$28..$Z$33,Z$11)*$A$4,0)," ")
17	@IF($A$7>=Z$11,Z$15-Z$16," ")
18	@IF($A$7>=Z$11,@ROUND($A$9*Z$17,0)," ")
19	@IF($A$7>=Z$11,Z$17-Z$18," ")
20	@IF($A$7>=Z$11,Z$16+Z$19," ")
21	@IF($A$7>=Z$11,1/(1+$A$10)^Z$11," ")
22	@IF($A$7>=Z$11,@ROUND(Z$20*Z$21,0)," ")

Figure 5-2C Formulae for NPWTAX Worksheet

	A	B	C	D	E	F	G
1	NPWTAXEX			G27			
4	$1,000,000	First Cost, FC					
5	$100,000	Additions to Working Capital, WC					
6	$100,000	Estimated Salvage Value, S					
7	6	Useful Life, N (Warning: MACRS Class < N < 26)					
8	5	MACRS Class (Enter 3,5,7,10,15, or 20)					
9	40.00%	Tax Rate (Enter, for example, 40% or .4)					
10	20.00%	MARR (Enter, for example, 15% or .15)					
12		YEAR 1	YEAR 2	YEAR 3	YEAR 4	YEAR 5	YEAR 6
13	Revenues	$600,000	$900,000	$800,000	$900,000	$600,000	$700,000
14	Operating Costs	$100,000	$200,000	$300,000	$400,000	$500,000	$600,000
15	Operating Income	$500,000	$700,000	$500,000	$500,000	$100,000	$100,000
16	Depreciation	$200,000	$320,000	$192,000	$115,200	$115,200	$57,600
17	Taxable Income	$300,000	$380,000	$308,000	$384,800	($15,200)	$42,400
18	Taxes	$120,000	$152,000	$123,200	$153,920	($6,080)	$16,960
19	Net Income	$180,000	$228,000	$184,800	$230,880	($9,120)	$25,440
20	Operating NCF	$380,000	$548,000	$376,800	$346,080	$106,080	$83,040
21	PW Factor	0.833333	0.694444	0.578704	0.482253	0.401878	0.334898
22	PW NCF	$316,667	$380,556	$218,056	$166,898	$42,631	$27,810
23	Total PW NCFs	$1,152,618					
24	PW A-T Salvage	$20,094					
25	PW Add. to WC	$66,510					
26	NPW	$106,202					
27	EAW	$31,936					

Figure 5-3 NPWTAX Example

An example calculation of NPW with taxes is given in Figure 5-3. The $1,000,000 first cost, $100,000 additions to working capital, $100,000 salvage value, 6 year useful life, 5 year MACRS class, 40.00% tax rate, and 20.00% MARR are entered in rows 4 through 10 of column A. The yearly revenues are entered in columns B through G of row 13. The yearly operating costs are entered in row 14. The NPW of $106,202 is calculated in cell B26. The EAW of $31,936 is in cell B27. Note that the name of the worksheet has been changed to NPWTAXEX in cell A1. The lower right hand print range label in cell D1 has also been changed to G27.

Sensitivity, Risk, and Inflation Analyses

Spreadsheet software is particularly well suited to *"what if"* or *sensitivity analysis, risk or scenario analysis, and inflation analysis.*

Sensitivity Analysis

Suppose you want to know how the NPW of the project in Figure 5-3 varies with changes in the MARR. This is often referred to as the *NPW profile.* All you have to do is enter a new data value for the MARR in cell A10, and the new NPW quickly appears in cell B26. Try this for MARR values of 0.10, 0.15, and 0.25. You should obtain NPWs of $420,876, $248,240, and -$12,133. This project is indeed sensitive to changes in the discount rate or MARR. A 50% decline in the MARR from 0.20 to 0.10 almost quadruples the NPW. An 25% increase in the MARR from 0.20 to 0.25 makes the project unprofitable. Varying the MARR is one type of risk analysis. That is, the MARR for a more risky project is higher than the MARR for a less risky project.

Sensitivity analysis can also be easily applied to the additions to working capital, the salvage value, and the tax rate.

Decreasing the required additions to working capital to zero in cell A5 increases NPW to $172,712 and increasing the additions to $200,000 decreases NPW to $39,692.

The NPWTAX model assumes that whatever additions to working capital that are required now, such as an increase in inventories, are released when the project is over. That is, the base case assumption for the project is that working capital is increased by $100,000 today but that this $100,000 is returned when the project ends in 6 years.

Decreasing the salvage value in cell A6 to zero decreases NPW to $86,108 and increasing the salvage value to $200,000 increases NPW to $126,296. Thus, the NPW of this project is more sensitive to additions to working capital than it is to the salvage value. *As a general rule, the longer the useful life of a project, the more likely is this result.*

It seems that almost every year there are some legislated tax changes. How sensitive is this project to changes in tax rates? A 25% increase in the tax rate to 0.50 in cell A9 reduces the NPW to $14,273 and a 50% increase in the tax rate to 0.60 changes the NPW to -$77,655. Thus, this project, as is true of most projects, is certainly sensitive to the tax rate. However, a 25% increase in rates is a rather massive change and this project would still be profitable.

Perhaps the most difficult data to forecast for a project are the yearly revenues. A 5% decrease in yearly revenues to $570,000 for year 1, $855,000 for year 2, $760,000 for year 3, $855,000 for year 4, $570,000 for year 5, and $665,000 for year 6 reduces NPW by slightly over 70% to $31,275. A 10% decrease in yearly revenues to $540,000 for year 1, $810,000 for year 2, $720,000 for year 3, $810,000 for year 4, $540,000 for year 5, and $630,000 for year 6 changes NPW to -$43,651. Thus, this project is highly sensitive to revenue projections. This is typical of most projects. *Because of this, we cannot stress enough the importance of reliable revenue estimates when performing an engineering economics analysis.*

In most industries, yearly operating costs can usually be more reliably estimated than revenues. However, there are some exceptions. Public utility revenues are typically regulated and may be known more reliably than operating costs, especially if a union contract negotiation looms.

In the example project, a 5% increase in yearly operating costs to $105,000 for year 1, $210,000 for year 2, $315,000 for year 3, $420,000 for year 4, $525,000 for year 5, and $630,000 for year 6 reduces NPW to $76,483. A 10% increase in yearly operating costs to $110,000 for year 1, $220,000 for year 2, $330,000 for year 3, $440,000 for year 4, $550,000 for year 5, and $660,000 for year 6 changes NPW to $46,765. Thus, this project is sensitive to cost projections.

Note, though, that the project is more sensitive to a specific percentage change in revenue projections than it is to the same percentage change in operating costs. *This is typical of most projects.*

Scenario Analysis

Since actual revenues often vary considerably from projected revenues, some type of *risk analysis* is frequently applied to revenues. *Scenario analysis* combined with the calculation of the *standard deviation* of NPW and the *coefficient of variation* of the NPW is a useful approach to risk analysis.

Suppose the *base case* yearly revenues in Figure 5-3 are the *most likely* scenario with a probability of 0.50. Further assume that a 10% increase in yearly revenues is the *most optimistic* scenario with a probability of 0.25 and that a 10% decrease in yearly revenues is the *most pessimistic* scenario with a probability of 0.25.

The NPW for the most optimistic scenario or outcome is \$256,055. This result is obtained by entering yearly revenues of \$660,000, \$990,000, \$880,000, \$990,000, \$660,000, and \$770,000 in cells B13 through G13 in place of the base case yearly revenues in the NPWTAX worksheet example in Figure 5-3. As noted above, the base case and 10% decline in revenue case result in NPWs of \$106,202 and -\$43,651, respectively.

The expected NPW, E(NPW), for the project is $\{[(0.25)\text{x}(\$256,055)] + [(0.50)\text{x}(\$106,202)] + [(0.25)\text{x}(-\$43,651)]\} = \$106,202$.

The standard deviation of the NPW, σ, is $\{[(0.25)\text{x}(\$256,055-\$106,202)^2] + [(0.50)\text{x}(\$106,202-\$106,202)^2] + [(0.25)\text{x}(-\$43,651-\$106,202)^2]\}^{0.5} = \$105,962$. The standard deviations of two projects are not really directly comparable if the projects have different first costs or additions to working capital.

The coefficient of variation, σ/E(NPW), is a measure of variability or risk per dollar of expected NPW. The larger the coefficient of variation, the riskier is the project. The coefficient of variation is useful for comparing projects of different sizes. The coefficient of variation for the project in Figure 5-3 is \$105,962 / \$106,202 = 0.9977. This is very close to 1, and *projects with coefficients of variation of 1 or higher are quite risky* as a NPW of 0 is within one standard deviation of the expected NPW. The high value of the coefficient of variation for this project is not surprising in view of the probability of 0.25 of having a substantially negative NPW.

Inflation Analysis

One factor that heavily influences revenue and operating cost projections is the rate of inflation. Revenues and costs, of course, can be subject to different rates of inflation. In regulated public utilities, for example, costs often increase before a rate hearing is held to allow increased revenues. Thus, for the practicing engineering economist, *no sensitivity analysis is complete without an inflation analysis.* (See the ESCALATE worksheet in Chapter 3 and the REGRESS worksheet in Appendix 3A.)

Replacement Analysis

The NPWTAX worksheet can be easily extended to do replacement analysis. Figure 5-4 presents this extended worksheet, named REPLACE, which calculates the *incremental* NPW of replacing an existing asset with a new asset.

Load your spreadsheet software and retrieve the file REPLACE.WKS from your working copy of the diskette that accompanies this book. There are four more lines in the REPLACE worksheet than there are in the NPWTAX worksheet. The estimated salvage value, which was entered in cell A6 of the NPWTAX worksheet, becomes the estimated salvage value of the new asset and is entered in cell A8 of the REPLACE worksheet. Cells A6 and A7 of the REPLACE worksheet require the entry of the current salvage value of the old asset and the terminal salvage value of the old asset, respectively. The REPLACE worksheet explicitly assumes that the old asset could be operated for N more years, where N is the useful economic life of the new asset.

The REPLACE worksheet has the abbreviation "Inc." added to the relevant labels in the range of cells A15..A31. This is to remind the user that, for example, the yearly revenues must be entered on an incremental basis, starting in cell B15. That is, one must calculate and enter the yearly revenues generated by the new asset less the yearly revenues generated by the old asset.

In some replacement projects, revenues are unchanged. It is operating costs that are reduced. In such cases, 0 would be entered for all the yearly revenues and yearly operating cost savings would be entered as negative values, starting with cell B16. The abbreviation "Inc." has also been added to the label for operating costs.

Replacement projects, of course, may involve a combination of revenue enhancement and cost savings. A worthwhile replacement project could even involve an increase in operating costs accompanied by sufficient revenue enhancement. In this age of increasing environmental concern, a replacement project may be mandated by government legislation. Such a project may have a negative incremental NPW compared to continued operation of the existing asset. If there are several alternatives that meet the environmental standard and have the same economic life, the one with the highest incremental NPW should be chosen. If the highest incremental NPW is negative, then the *alternative of abandoning the operation of the non-conforming asset must be considered.* The NPW of the best replacement project should then be calculated by the NPWTAX worksheet to confirm that its NPW is positive. Otherwise, the existing non-conforming asset must be scrapped instead of replaced.

Depreciation on the new asset is calculated in the REPLACE worksheet exactly like it was in the NPWTAX worksheet. That is, the REPLACE worksheet is restricted to new assets in MACRS 3, 5, 7, 10, 15, and 20 year classes, and the useful economic life must be greater than the MACRS class life but less than 26 years.

Depreciation on the old asset is not calculated by the REPLACE worksheet. The reason for this is that there is such an incredible number of possibilities for the old asset, depending on when it was purchased. Usually, a depreciation schedule already exists for the old asset. Thus, the remaining yearly depreciation can be transferred from this schedule to row 19 of the REPLACE worksheet, starting with column B.

	A	B	C	D
4		First Cost, FC, of New Asset		
5		Additions to Working Capital, WC		
6		Current Salvage Value, Old Asset		
7		Estimated Salvage Value at Time N, Old Asset		
8		Estimated Salvage Value at Time N, New Asset		
9		Useful Life, N (Warning: MACRS Class < N < 26)		
10		MACRS Class (Enter 3,5,7,10,15, or 20)		
11		Tax Rate (Enter, for example, 40% or .4)		
12		MARR (Enter, for example, 15% or .15)		
15	Inc. Revenues			
16	Inc. Operating Costs			
17	Inc. Operating Income			
18	Depreciation, New Asset			
19	Depreciation, Old Asset			
20	Inc. Depreciation			
21	Inc. Taxable Income			
22	Inc. Taxes			
23	Inc. Net Income			
24	Inc. Operating NCF			
25	PW Factor			
26	PW Inc. NCF			
27	Total PW Inc. NCFs			
28	PW Inc. A-T Salvage			
29	PW Add. to WC			
30	Inc. NPW			
31	Inc. EAW			

Figure 5-4 REPLACE Worksheet

	B
13	1+A$13
14	@IF($A$9>=B$13," YEAR 1"," ")
17	@IF($A$9>=B$13,B$15-B$16," ")
18	@IF($A$9>=B$13,@ROUND(@VLOOKUP($A$10,$A$32..$Z$37,B$13)*$A$4,0)," ")
20	@IF($A$9>=B$13,B$18-B$19," ")
21	@IF($A$9>=B$13,B$17-B$20," ")
22	@IF($A$9>=B$13,@ROUND($A$11*B$21,0)," ")
23	@IF($A$9>=B$13,B$21-B$22," ")
24	@IF($A$9>=B$13,B$20+B$23," ")
25	@IF($A$9>=B$13,1/(1+$A$12) ^ B$13," ")
26	@IF($A$9>=B$13,@ROUND(B$24*B$25,0)," ")
27	@IF($A$9>=$B$13,@SUM($B$26..$Z$26)," ")
28	@IF($A$9>=$B$13,@ROUND($A$8-$A$7)*(1-$A$11)/(1+$A$12) ^ $A$9,0)," ")
29	@IF($A$9>=$B$13,@ROUND($A$5*(1-1/(1+$A$12) ^ $A$9,0)," ")
30	@IF($A$9>=$B$13,-$A$4+$A$6+$B$27+$B$28-$B$29," ")
31	@IF($A$9>=$B$13,($B$30*$A$12)/(1-1/(1+$A$12) ^ $A$9)," ")

Figure 5-5 Formulae for REPLACE Worksheet

Since the structure of the two worksheets is so similar, Figure 5-5 illustrates the essence of these changes by presenting the formulae for column B. As previously noted, the REPLACE worksheet differs from the NPWTAX worksheet by the addition of 4 rows. Many of the required changes in the formulae occurred automatically when the new rows were inserted in the NPWTAX worksheet to create the REPLACE worksheet.

Compare this to the formulae for column B for the NPWTAX worksheet in Figure 5-2A. Cell B21 was modified to deduct the yearly incremental depreciation in the relevant cell in row 20, rather than the yearly depreciation on the new asset in row 18. (Cells C21 through Z21 were, of course, similarly modified.) Similarly, in cell B24 and cells C24 through Z24, incremental depreciation (instead of depreciation on the new asset) is added to incremental net income to get incremental operating net cash flow. (Although the depreciation rates are not shown, they are in rows 32 through 37 instead of rows 28 through 33.)

You should also pay particular attention to the formula in cell B30 for the calculation of incremental NPW. Note that the current salvage value of the old asset has been added to this formula. Generally there is also a tax consequence to selling the old asset. We have implicitly assumed that the old asset is being sold for its remaining book value. This is rarely true.

You may wish to revise the REPLACE worksheet to account for these tax effects. First, globally unprotect the worksheet by typing ⌷ⓌⒼⓅⒹ . Then, insert another row by moving the cursor to cell A6 and pressing ⌷ⓌⒾⓇⒺⁿᵗᵉʳ . The formulae in REPLACE will be automatically adjusted for the addition of this row. Next, move the cursor to cell B6, type in the label **Book Value, Old Asset**, and press Ⓔⁿᵗᵉʳ . Now move the cursor to cell B31 and edit the incremental NPW formula. (Remember, the addition of a row moved the incremental NPW calculation from B30 to B31.) To edit B31 for the tax effects of the difference between the current salvage value and book value of the old asset, press the edit key Ⓕ②, type **-(\$A\$7-\$A\$6)*\$A\$12**, and press ⏎. Restore global protection by typing: ⌷ⓌⒼⓅⒺ .

What you have just done is subtracted the taxes incurred from the sale of the old asset. That is, **(\$A\$7-\$A\$6)** represents the difference between the salvage and book values of the old asset and cell A12 now contains the tax rate after the new row 6 was inserted in the worksheet.

If you can defer these taxes until the end of year 1, then you should discount these taxes for one period by typing **-(\$A\$7-\$A\$6)*\$A\$12/(1+\$A\$13)**, instead of **-(\$A\$7-\$A\$6)*\$A\$12**. You should also replace the label **Z31** in cell D1 with the label **Z32** to reflect that the range of the visible worksheet extends from A1..Z32. Hidden from view, of course, are the MACRS depreciation rates, which are now in rows 33 through 38 starting with column B.

Figure 5-6 gives an example of the unmodified REPLACE worksheet. In this example an asset with an expected life of 15 years was purchased $1,500,000 and is being replaced with a new asset in the MACRS 3 year class that will last 4 years.

	A	B	C	D	E
4	$1,000,000	First Cost, FC, of New Asset			
5	$0	Additions to Working Capital, WC			
6	$400,000	Current Salvage Value, Old Asset			
7	$0	Estimated Salvage Value at Time N, Old Asset			
8	$300,000	Estimated Salvage Value at Time N, New Asset			
9	4	Useful Life, N (Warning: MACRS Class < N < 26)			
10	3	MACRS Class (Enter 3,5,7,10,15, or 20)			
11	40.00%	Tax Rate (Enter, for example, 40% or .4)			
12	10.00%	MARR (Enter, for example, 15% or .15)			
14		YEAR 1	YEAR 2	YEAR 3	YEAR 4
15	Inc. Revenues	$0	$0	$0	$0
16	Inc. Operating Costs	($250,000)	($250,000)	($250,000)	($250,000)
17	Inc. Operating Income	$250,000	$250,000	$250,000	$250,000
18	Depreciation, New Asset	$333,300	$444,500	$148,100	$74,100
19	Depreciation, Old Asset	$100,000	$100,000	$100,000	$100,000
20	Inc. Depreciation	$233,300	$344,500	$48,100	($25,900)
21	Taxable Income	$16,700	($94,500)	$201,900	$275,900
22	Inc. Taxes	$6,680	($37,800)	$80,760	$110,360
23	Inc. Net Income	$10,020	($56,700)	$121,140	$165,540
24	Inc. Operating NCF	$243,320	$287,800	$169,240	$139,640
25	Inc. PW Factor	0.909091	0.826446	0.751315	0.683013
26	PW Inc. NCF	$221,200	$237,851	$127,153	$95,376
27	Total PW Inc. NCFs	$681,580			
28	PW Inc. A-T Salvage	$122,942			
29	PW Add. to WC	$0			
30	Inc. NPW	$204,522			
31	Inc. EAW	$64,521			

Figure 5-6 REPLACE Example

Since the old asset was purchased before the Economic Recovery Act of 1981, straight line depreciation was permitted and was used by the firm. (Under the old rules, alternative depreciation methods, such as declining balance and sum-of-the-years digits, were allowed and frequently used in the early years of an asset's life since higher depreciation write-offs increased cash flows. However, these accelerated methods usually became less advantageous as the asset became older so most firms changed to the straight line method in the later years of an asset's life). Thus, the current book value of the old asset is $400,000, which is also its current salvage value, and the permitted depreciation for the next 4 years would be $100,000 per year if it is continued in operation.

The new asset will not generate additional revenues but will provide cost savings of $250,000 per year. The MARR is 10%, the tax rate is 40%, and the salvage values of the new and old assets in 4 years are $300,000 and $0, respectively. The purchase price of the new asset is $1,000,000 and no changes to working capital are required.

Should the old asset be replaced? According to cell B30 in Figure 5-6, the incremental NPW is $204,522. Hence, the old asset should be replaced because the incremental NPW of replacement is positive.

Replacement with Unequal Lives

The REPLACE worksheet assumes that the old asset can be operated for as long as the new asset. In practice this does not always occur. In such a case, instead of modifying the REPLACE worksheet, we would use the NPWTAX worksheet to calculate the new asset's EAW. Next, we would modify the NPWTAX worksheet to calculate the first cost of the old asset as its current salvage value less any tax effects from scrapping this asset. Thus, the first cost becomes the opportunity cost of keeping the old asset. Depreciation for the old asset, of course, must *not* be based on this opportunity cost, but on the schedule which already exists for the old asset. Then, we would use this modified NPWTAX worksheet to calculate EAW of the old asset. Finally, the EAW of the new asset must be compared with the EAW of the old asset. The asset with the higher positive EAW should be selected. *If both assets have negative EAW, the old asset should simply be scrapped rather than replaced.* Such an approach should also be considered when both the old and new assets have equal lives if it is suspected that both assets *may* have negative NPW.

Appendix 5A

B.C. MINES: A CASE STUDY

To demonstrate the estimation and use of cash flows, we study here the case of B.C. Mines Ltd., a Canadian subsidiary of U.S. Mines Inc. B.C. Mines is considering opening a new open pit mine located in a remote northern part of the western province of British Columbia. The mining company has leased the mineral rights to a tract of Indian land for a 5% net smelter return royalty. B.C. Mines uses a MARR of 20%.

Feasibility

The feasibility study has indicated the following key parameters: ore reserves are 4,860,000 tons with a grading of 1.45 percent; mill recovery is 90 percent; production is 1,500 tons per day or 540,000 tons per year; operating costs are $10 per ton milled; capital costs are $15,000,000; and the economic life is nine years (4,860,000 tons / 540,000 tons).

Operations

Since the construction period is twelve months, or all of Year 0, the startup of commercial production is January 1 of Year 1. For simplicity, assume that the capital costs are payable on January 1 of Year 1, which is denoted as time 0.

The recovered mineral in any year is:

Tons of ore x grade x 2,000 pounds x % recovery = 540,000 x (1.45/ 100) x 2,000 x (90/100) = 14,094,000 pounds.

The net smelter return is the market price of the commodity less the off-site costs of concentrate freight to the smelter, smelting charges, anode freight charges, refining costs, concentrate losses, etc. If the mineral price is $1.10 per pound and the total deductions are 30 cents per pound, the net smelter return will be 80 cents per pound. Thus, the annual net smelter return or sales revenue is 14,094,000 x ($1.10 - 0.30) = $11,275,000.

The annual royalties are 5% x $11,275,000 or $563,750 per year and are payable whether the operation of the mine is profitable or not.

The operating costs of $10 per ton include mining the ore, stripping waste, milling, plant engineering, and maintenance, as well as general and administrative costs. Plant engineering costs are all general charges for plant operation that are not charged specifically to mining or milling, such as yard lighting, snow removal, plant heating, etc. The annual operating cost is 540,000 x $10 = $5,400,000.

The operating profit is the net smelter return minus operating cost and royalties:

	Per Year
Net Smelter Return	$11,275,000
Less NSR Royalty	563,750
Project Revenues	$10,711,250
Less Operating Cost	5,400,000
Operating Profit	$5,311,250

Working Capital

The working capital expenditure is the amount required to pay operating costs from the start of production until the initial sales revenue is received. Generally in mining operations 3 to 4 months of operating costs must be covered. If we opt for 3 months, we will need (3/12) x 540,000 x $10 = $1,350,000. This working capital will be constant over the life of the mine, since there will always be a delay between production and the receipt of income from that production. The working capital will be recovered shortly after the mine closes, but in engineering economics work we usually assume that it will be recovered at the end of the economic life of the mine, in Year 9.

First Cost

Capital expenditures include the buildings, machinery, furnishings, dams, mobile equipment, head frames, and other such tangible items needed to put the new mine into operation. These capital assets are categorized as either processing assets ($5,625,000) or as mining and service assets ($9,375,000), and are subject to different depreciation treatment for tax purposes.

Taxation

The Canadian *Capital Cost Allowance* (CCA) or tax depreciation system is considered unique. Every fixed asset belongs to an asset class and each class is accounted for separately. The basis for this accounting is the capital cost, which generally means the full cost to the taxpayer of acquiring the asset, including legal, accounting, engineering, and other costs in addition to the purchase price. Each class has its own CCA rate. There are two types of CCA rates: declining balance (used for almost all tangible assets) and straight-line (used for most intangible assets). The CCA rules are complex and change frequently. As of 1991, the processing assets are in class 39 (with a CCA rate of 25%) and the mining and service assets are in class 41 (which also has a CCA rate of 25%). Since the two rates, which were historically different, now are the same, and since B. C. Mines expects to have assets in both classes well beyond the life of this proposed project, we can combine the project assets (which total $15,000,000) for purposes of calculating the yearly project CCA. Expected salvage value for the combined project assets is $1,250,000.

If the rates had still been different, this potentially could affect the design of the project. (In actual fact the number of asset classes could be even more than two. If the dam and buildings are considered permanent structures, they would go in class 1 with a CCA rate of 4%.) Suppose, for example, that the mining and service assets had a higher CCA rate. Then the present worth of project cash flows could be increased if the project design can be altered to increase the amount of mining and service assets while effecting a corresponding reduction in processing assets.

That is, *a faster write-off of assets reduces taxable income in the early years, hence, taxes. This increases cash flows in the early years because operating net cash flows are simply operating income less taxes, which is the same as net income plus depreciation.*

The federal and provincial governments of Canada levy income taxes on both individuals and corporations who are residents of Canada. They also tax the Canadian-source income of nonresidents. Canadian residents are taxed by Canada on *all* their income, regardless of the source, inside or outside Canada. Thus, taxes will have to be paid to the Province of British Columbia as well as to the federal government of Canada. Income taxes to both the provincial and federal governments are applied to operating profit after deducting CCA. The tax rules of the provincial and federal governments are not necessarily the same. In particular, there is a provincial mining tax in addition to the provincial income tax. There are also various resource and depletion allowances that may differ federally and provincially.

The purpose of this case, however, is to provide a general illustration of the Canadian CCA system. Therefore, we shall ignore the tax intricacies peculiar to the mining industry. Canadian tax law is important to many U.S. companies, as they have subsidiaries in Canada. American investment in Canada has been further fostered by the U.S-Canadian Free Trade Agreement. In this case, the combined federal and provincial tax rate is 40%.

Worksheet

Load your spreadsheet software and retrieve the NPWCNTAX.WKS worksheet file. This worksheet is virtually identical to the NPWTAX worksheet shown in Figure 5-1.

The principal difference between the two worksheets is in data entry. Note that the CCA rate (not the CCA class) is entered in cell A8. (The MACRS class was entered in cell A8 of NPWTAX worksheet.) There are also changes to the calculation of tax depreciation (labelled CCA) in row 16.

The formula in cell B16 of Figure 5A-1A calculates the CCA for Year 1. Just like the U.S., Canada follows the half-year convention. Thus, the permitted CCA rate in the first year is only half the normal rate for the class.

	B
16	@IF($A$7>=B$11,@ROUND(0.5*$A$4*$A$8,0),"")
24	@IF($A$7>=$B$11,@ROUND($A$6+($A$4-@SUM($B$16..$Z$16))-$A$6)*$A$8*$A$9/($A$8+$A$10))/(1+$A$10)^$A$7,0)," ")

Figure 5A-1A Formulae for NPWCNTAX Worksheet

The formula in cell C16 of Figure 5A-1B subtracts the CCA taken in Year 1 from the First Cost to get the Remaining Book Value (Undepreciated Capital Cost) of the assets. The Remaining Book Value is then multiplied by the normal CCA rate to get the permitted CCA for Year 2.

	C
16	@IF($A$7>=C$11,@ROUND(($A$4-@SUM($B$16..B$16))*$A$8,0),"")

Figure 5A-1B Formulae for NPWCNTAX Worksheet

The formulae for CCA in Years 3 through 25 in cells D16..Z16 are identical to the formula in cell C16, except that the function @SUM($B$16..B$16) is replaced by @SUM($B16..C$16) through @SUM($B16..Y$16), respectively.

The only other formula change is in cell B24 of Figure 5-1A. This change reflects the major difference between Canadian and United States tax law. In the U.S., the MACRS tables are initially based on the declining balance depreciation method, but then switch over to the straight line depreciation method when the straight line amount exceeds the declining balance amount.

Thus, the first cost of the asset is eventually fully depreciated. In Canada, the switch to straight line is not allowed. *Hence, in Canada, an asset is **never** fully depreciated.* Because assets are never fully depreciated in Canada, the formula in cell B24 reflects the fact that the difference between the Remaining Book Value at time N and the salvage value continues to generate CCA tax savings in Years N+1 through ∞.

It can be shown that the future worth at time N of the CCA tax savings for years N+1 through ∞ is the difference between the Remaining Book Value and the salvage value multiplied by the product of the tax and depreciation rates and then divided by the sum of the MARR and depreciation rates. If the salvage value exceeds the Remaining Book Value (but not the Remaining Book Value of all assets in the class and not the asset's original cost base), then CCA tax savings from other class assets are correspondingly reduced in Years N+1 through ∞.

Figures 5A-2A and B show the B.C. Mines data in the NPWCNTAX worksheet (renamed BCMINES). The project NPW is $1,299. This is a very small NPW for a $15,000,000 investment. This slightly positive NPW is heavily dependent on a positive salvage value of $1,250,000. *Salvage values are notoriously difficult to estimate. Thus, we advise against proceeding with this project even though its NPW is positive.*

	A	B	C	D	E
1	BCMINES				J27
4	$15,000,000	First Cost, FC			
5	$1,350,000	Additions to Working Capital, WC			
6	$1,250,000	Estimated Salvage Value, S			
7	9	Useful Life, N (Warning: N < 26)			
8	25.00%	CCA Rates (Enter, for example, 30% or .3)			
9	40.00%	Tax Rate (Enter, for example, 40% or .4)			
10	20.00%	MARR (Enter, for example, 15% or .15)			
12		YEAR 1	YEAR 2	YEAR 3	YEAR 4
13	Revenues	$10,711,250	$10,711,250	$10,711,250	$10,711,250
14	Operating Costs	$5,400,000	$5,400,000	$5,400,000	$5,400,000
15	Operating Income	$5,311,250	$5,311,250	$5,311,250	$5,311,250
16	CCA	$1,875,000	$3,281,250	$2,460,938	$1,845,703
17	Taxable Income	$3,436,000	$2,030,000	$2,850,312	$3,465,547
18	Taxes	$1,374,000	$812,000	$1,140,125	$1,386,219
19	Net Income	$2,061,750	$1,218,000	$1,171,125	$2,079,328
20	Operating NCF	$3,936,750	$4,499,250	$4,171,125	$3,925,031
21	PW Factor	0.833333	0.694444	0.578704	0.482253
22	PW NCF	$3,280,625	$3,124,479	$2,413,845	$1,892,858
23	Total PW NCFs	$15,844,646			
24	PW A-T Salvage	$245,014			
25	PW Add. to WC	$1,088,361			
26	NPW	$1,299			
27	EAW	$322			

Figure 5A-2A B.C. Mines Project

	P	G	H	I	J
12	YEAR 5	YEAR 6	YEAR 7	YEAR 8	YEAR 9
13	$10,711,250	$10,711,250	$10,711,250	$10,711,250	$10,711,250
14	$5,400,000	$5,400,000	$5,400,000	$5,400,000	$5,400,000
15	$5,311,250	$5,311,250	$5,311,250	$5,311,250	$5,311,250
16	$1,384,277	$1,709,217	$1,813,038	$1,890,903	$1,949,302
17	$3,926,973	$4,273,042	$4,532,594	$4,727,258	$4,873,256
18	$1,570,789	$1,709,217	$1,813,038	$1,890,903	$1,949,302
19	$2,356,184	$2,563,825	$2,719,556	$2,836,355	$2,923,954
20	$3,740,461	$3,602,033	$3,498,212	$3,420,347	$3,361,948
21	0.401878	0.334898	0.279082	0.232568	0.193807
22	$1,503,207	$1,206,314	$976,287	$795,463	$651,568

Figure 5A-2B B.C. Mines Project

CCA vs. MACRS

Figure 5A-3 presents a worksheet in which the U.S. data from the NPWTAX example shown in Figure 5-3 have been entered into the NPWCNTAX worksheet, except that a declining balance depreciation rate of 40% is entered in cell A8.

Because of the half-year convention, the rate used in calculations in Year 1 is only 20%. However, in Year 2, the rate is 40%. This is equivalent to applying $(1-.2)(.4) = 0.32000$ to the First Cost because the Remaining Book Value at the beginning of Year 2 is (First Cost) x $(1-.2)$, where .2 is the first year depreciation rate. Similarly, in Year 3 the full 40% is applied to the Remaining Book Value at the beginning of Year 3, which is equivalent to applying $(1-.2)(1-.4)(.4) = .19200$ to the First Cost. In Year 4, the equivalent First Cost depreciation rate is $(1-.2)(1-.4)^2(.4) = 0.15520$. *Thus, the operating NCFs for Years 1-4 are the same in Figure 5-9 as in Figure 5-3 because the effective First Cost depreciation rates in Canada of 0.20000, 0.32000, 0.19200, and 0.11520, respectively, match the MACRS rates in cells B29..E29 in Figure 5-2B.*

	A	B	C	D	E	F	G
1	TAXCOMP			G27			
4	$1,000,000	First Cost, FC					
5	$100,000	Additions to Working Capital, WC					
6	$100,000	Estimated Salvage Value, S					
7	6	Useful Life, N (Warning: N < 26)					
8	40.00%	CCA rate (Enter, for example, 30% or .3)					
9	0.4000	Tax Rate (Enter, for example, 40% or .4)					
10	0.2000	MARR (Enter, for example, 15% or .15)					
12		YEAR 1	YEAR 2	YEAR 3	YEAR 4	YEAR 5	YEAR 6
13	Revenues	$600,000	$900,000	$800,000	$900,000	$600,000	$700,000
14	Operating Costs	$100,000	$200,000	$300,000	$400,000	$500,000	$600,000
15	Operating Income	$500,000	$700,000	$500,000	$500,000	$100,000	$100,000
16	CCA	$200,000	$320,000	$192,000	$115,200	$69,120	$41,472
17	Taxable Income	$300,000	$380,000	$308,000	$384,800	$30,880	$58,528
18	Taxes	$120,000	$152,000	$123,200	$153,920	$12,352	$23,411
19	Net Income	$180,000	$228,000	$184,800	$230,880	$18,528	$35,117
20	Operating NCF	$380,000	$548,000	$376,800	$346,080	$87,648	$76,589
21	PW Factor	0.833333	0.694444	0.578704	0.482253	0.401878	0.334898
22	PW NCF	$316,667	$380,556	$218,056	$166,898	$35,224	$25,650
23	Total PW NCFs	$1,143,051					
24	PW A-T Salvage	$30,115					
25	PW Add. to WC	$66,510					
26	NPW	$106,656					
27	EAW	$32,072					

Figure 5A-3 Comparison of U.S. and Canadian Taxation

In Year 5, however, the equivalent First Cost depreciation rate is $(1-.2)(1-.4)^3$ $(.4) = 0.04608$ in Canada, which is less than the MACRS rate of 0.11520. This is because MACRS has switched to straight line and the Year 5 MACRS rate is the same as the Year 4 rate. (The Year 6 MACRS rate is half the Year 5 rate, which is an artifact of the half-year convention in Year 1.)

Note that the NPW of $106,202 from the U.S. investment in Figure 5-3 is less than the NPW of $106,656 for the comparable Canadian investment. *Thus, in this example, the Canadian investment is preferable.*

Initially, this result may seem surprising because the U.S. MACRS leads to higher operating NCFs in Years 5 and 6. Under MACRS, however, the difference between the salvage value and the Remaining Book Value (0 in this example) is considered a recapture of excess depreciation already taken. This triggers an immediate tax liability at ordinary income tax rates on the recapture.

In contrast, the CCA system triggers a recapture in Years N+1 through ∞ on the difference between the salvage value and the Remaining Book Value at the beginning of Year N+1.

Thus, it appears that country investment preference depends heavily on the salvage value. Performing sensitivity analysis on the salvage value in our example reveals that a salvage value of $89,820 yields an NPW of $104,156 in both countries. *U.S. investment is preferable for lower salvage values and Canadian investment is preferable for higher salvage values.*

Note that the NPWTAX worksheet implicitly assumes that the asset is sold on December 31 of Year N; whereas, the NPWCNTAX worksheet implicitly assumes that the asset is sold on January 1 of Year N+1. This is consistent with American and Canadian textbooks that the authors have used. A fairer comparison of the two taxation systems would assume that the asset is sold on January 1 in both worksheets. The required modification to NPWTAX is to replace **$A$9** in the formula in cell B24 with **$A$9/(1+$A$10)**. That is, taxes on the recapture are paid at the end of Year N+1, instead of at the end of Year N. *This does not change the qualitative nature of our results but does extend the range of preference for the U.S. investment.* With a salvage value of $100,000, the NPW of the U.S. investment becomes $108,435. With a salvage value of $179,700, the NPW in both countries is $126,229.

Chapter 6
FINANCIAL DATA ANALYSIS

Engineers frequently need to know about the financial status of a particular company, for example, when they decide whether or not to subcontract some work or grant credit. In addition, engineers as well as managers may want to know how their firm compares with its own previous performance, its competitors, or with some absolute standard. In all these situations, decisions are usually based on financial ratio analysis.

Calculating one ratio is a simple process, but calculating many is both tedious and time consuming. A personal computer can speed up this process, while also diminishing the probability of computational errors.

A sample analysis of a Canadian high tech firm, Gandalf Technologies, whose securities are traded in both the U.S. and Canada, is used to demonstrate how a worksheet model can be used for financial data analysis.

A model, like the one described here, can be used over and over simply by changing the data. This capability also makes it easy to analyze "what if" situations by varying one or more items to see what happens. This worksheet model, called RATIOS, is included on the diskette which accompanies this book.

Annual Reports

The main source of financial information is the firm's annual report. The financial statements (balance sheet, income statement, statement of retained earnings, and statement of cash flows) included in the annual report are audited, thus adding to their credibility.

Annual reports usually also include a qualitative, verbal statement by management about the latest year's operations. A managerial forecast of future operations is often included.

The financial statements in these annual reports are prepared in accordance with generally accepted accounting principles (GAAP). However, in some industries, departures from GAAP are required by law or by the appropriate regulatory agencies. Accounting data are based on historical cost and allocation procedures, *not* cash flows.

Financial Ratios

Financial ratios are used to show relationships between and among the financial statement numbers. Ratios are usually categorized into the following four categories: liquidity, management, debt, and profitability. Some of the most commonly used ratios are discussed below.

Liquidity Ratios

A very important concern is liquidity − the ability to turn assets into cash without suffering undue losses. The key question is whether the firm will be able to pay its bills as they come due. Without sufficient liquidity in the short run, there will be no long run profits. A comprehensive analysis of liquidity obviously requires cash budgets. Since outsiders do not have access to the firm's cash budget information, they have to use what is available.

Ratio analysis of financial statement items gives a quickly calculated and easy to use measure of liquidity, albeit a rough one.

The *current ratio* is computed by dividing current assets by current liabilities. Current assets include cash, marketable securities, accounts receivable, and inventories. Current liabilities include accounts payable, notes payable, the current portion of long term debts that is payable in the next financial reporting period, and accrued expenses including taxes.

The benchmark current ratio for all firms is two to one. This would indicate that the firm has twice as much invested in current assets as it owes in current liabilities. Therefore, it is very likely that it would be able to pay its short-term debts as they come due, unless they are all due at the same time.

The rationale for the use of the current ratio is that it can serve as an early warning indicator of financial distress. If a company is getting into financial difficulty, it will probably try to borrow more and pay its bills more slowly. Thus, if current liabilities are growing faster than current assets, the current ratio will fall, and this declining trend indicates probable trouble.

The current ratio is the most widely used financial ratio. The current ratio has a defect, however, since the liquidity of its components is not always the same.

Although current liabilities are all short-term debts which must be paid soon, the current assets are not so homogeneous. Inventories are usually much less liquid than the other items. Cash on hand and temporary investments are very liquid. Accounts receivable, if paid soon, will be converted into cash. However, inventories must first be sold, creating accounts receivable in most cases, and this causes delay. Of course, it is not certain either that the sales of inventory can be made at all.

To remedy this defect, the *quick ratio* (or acid test ratio) is also used. The quick ratio is calculated by first deducting inventories from the current assets and then dividing the remaining figure by current liabilities. *The benchmark value for the quick ratio is one to one.*

The rationale for using the quick ratio is as follows: If accounts receivable can be collected as scheduled, a firm may be able to pay off its current liabilities with the cash on hand and the cash expected from accounts receivable without selling any of its inventory. This reasoning suggests that the quick ratio should always be at least one to one, unless it is virtually certain that some sales will be made. Thus, one can see with the quick ratio whether a firm is likely to be in trouble soon or not.

Both the current and quick ratios are calculated from the account items in the balance sheet. They are necessarily based on the data as of a specific date, which may be just before a large amount of short term debts must be paid. All of the balance sheet ratios are helpful in evaluating the situation of a firm over time by looking at the trend in these ratios, but they can provide only a rough idea of the firm's liquidity because they are *not based on expected cash flows*.

Management Ratios

The management ratios are intended to indicate how effectively the firm is managing its assets. This is done by considering whether the amount of each type of asset reported on the balance sheet appears to be reasonable for the level of operations of the firm. If assets are too low, then operations will not be efficient. If assets are too high, then interest costs are likely to be too high and profits will suffer.

The *inventory utilization ratio* (also called the inventory turnover ratio) is calculated by dividing sales by inventory.

Analysts outside the firm being analyzed usually use sales in the numerator of this ratio because the sales figure is easy to obtain. Most of the external reporting agencies, such as Dun and Bradstreet, use sales in the numerator of their ratios. Thus, if comparisons with industry averages are desired, we have to calculate ratios in the same way as the averages. Sales figures are normally stated at market prices. However, inventories are usually carried on the balance sheet at cost or market value, whichever is lower. Hence, it would be more appropriate to use cost of sales in the numerator rather than sales. Internal analysts usually have access to the necessary data. Thus, they can calculate the ratio both ways. However, they do tend to prefer to use the cost of sales.

There is a problem with this ratio. Sales occur during the year, but inventory is listed on the balance sheet as of a particular date. Therefore, it might be more appropriate to use average inventory for the denominator of this ratio. This is especially true if the firm's business is subject to significant seasonal fluctuations.

For most firms, the majority of sales are credit sales. Hence, collecting accounts receivable is of paramount importance. The *average collection period*, which can be used to appraise the effectiveness of the collection activities, is calculated by dividing average daily sales into accounts receivable to find the number of days sales tied up in receivables. This is defined as the average collection period (ACP) because it indicates the average length of time the firm must wait before actually getting the cash after making a sale. The ACP should be compared with the firm's credit terms. For example, if the terms of sale were net 30 days, any ACP greater than 30 would indicate that the firm's customers were not paying their bills on time.

This might indicate either a problem with credit policy (some customers are getting credit who should not) or it might just indicate that stronger steps need to be taken to accelerate collections.

An alternative indicator is the *receivables turnover ratio* which is calculated by dividing sales by accounts receivable. If the firm's credit terms are net 30 days, we would expect that accounts receivable on the firm's balance sheet would be only one twelfth of the sales on the firm's income statement. This would mean that the receivables are turning over twelve times a year.

The *fixed asset utilization ratio* (also called the fixed asset turnover ratio), which is calculated by dividing sales by fixed assets, serves as an indicator of the degree of utilization of plant and equipment.

A similar ratio, the *total asset utilization ratio* (or total asset turnover ratio), indicates the degree to which all of the assets are being utilized. It is calculated by dividing sales by total assets.

Debt Ratios

The extent to which a firm uses debt financing, or financial leverage, is important to creditors, shareholders, employees, suppliers, and the government since they would all suffer if the firm goes out of business as it might if it is unable to pay interest on its debts as scheduled.

If the firm can employ borrowed funds profitably, any return on the investment that exceeds the fixed cost of the debt will benefit the shareholders since profits will be enhanced. Also, raising large amounts of funds through debt allows the owners to control the firm with a limited investment of their own funds.

Creditors expect that owners' equity funds will provide a margin of safety if the expected returns do not materialize. If the owners have contributed only a very small portion of the total financing, then the risks of the firm will be borne by the creditors. If the creditors feel they may not be adequately compensated for risk by the interest return they expect, then they will not provide the funds.

Financial leverage can be evaluated either (1) by examining balance sheet ratios to see how much of the total funds have been borrowed or (2) by examining income statement ratios to see how many times the required interest payments (or the fixed charges for debt service payments, sinking fund payments, and lease payments) are covered by the operating profits of the firm. Both of these methods give complementary indications of the margin of safety which protects the firm's creditors. Financial analysts generally investigate both types of ratios.

The *debt ratio*, calculated by dividing total debt by total assets, measures the percentage of the total funds provided by the creditors. This ratio is also called the ratio of total debt to total assets. (Total assets is used in this ratio because the balance sheet is always in balance, so total assets equals the total of liabilities and capital.) Creditors naturally prefer low debt ratios because this gives them better protection against risk, while owners prefer higher leverage with its attendant possibility of higher profits.

The *times interest earned ratio* (frequently abbreviated as TIE), is calculated by dividing earnings before interest and taxes (EBIT) by the interest charges which must be paid. The TIE shows the extent to which earnings can decline without affecting the firm's ability to pay the interest it owes.

While interest is usually the major fixed charge, there may be others, such as sinking funds or leases.

Thus, in these cases, the *fixed charge coverage ratio* (abbreviated as FCC) is needed. The fixed charge coverage ratio is calculated in a manner analogous to the TIE, by adding the lease payments to the numerator and both sinking fund payments and lease payments to the denominator. The sinking fund payments have to be adjusted upward by dividing them by the term (1-T), where T is the marginal combined federal and state tax rate of the firm. This asymmetrical treatment is necessary because both interest and lease payments are tax deductible but sinking fund payments are not. The FCC recognizes that failure to make any of the required payments can result in bankruptcy.

Profitability Ratios

The preceding ratios are useful for evaluating the way the firm is operating in particular aspects of its activities. The profitability ratios show the combined effects of liquidity, asset management, and financial leverage.

The *margin ratio* is calculated by dividing net income by sales. This gives the profit per dollar of sales. Both of these data inputs come from the income statement.

Profitability, however, is more commonly measured by the *return on assets* (ROA). The ROA is calculated by dividing net income by total assets. This ratio indicates the profit generated by each dollar of assets, but it ignores the financial leverage, which might have been used to acquire the assets. Another way of calculating ROA is by multiplying the margin ratio by the total asset utilization ratio.

A better measure is provided by the *return on equity* (ROE). The ROE is calculated by dividing net income available to the common shareholders by the common equity shown on the balance sheet.

For those companies that use preferred stock, it is necessary to deduct the value of the preferred shares from the total shareholders equity to get the denominator of this ratio.

Both the ROA and ROE ratios can also be improved by using the average assets or the average equity in their denominators. Analysts who use publicly available comparative data need to check exactly how such ratios have been calculated if valid comparisons are desired.

The *earning power ratio* is calculated by dividing earnings before interest and taxes (EBIT) by total assets. Because earnings are generated over time, and assets are shown on the balance sheet as of a particular date, it is more appropriate to use average total assets in the denominator. This requires information from two balance sheets.

Limitations of Financial Data

Financial ratios can provide quite a bit of useful information concerning the operations of a firm and its financial situation. However, there are some limitations that require careful analysis. Some of the more important of these are discussed below.

Diversification

Many large firms operate in many different industries at the same time because they are so diversified. *It is difficult to develop meaningful comparative data for such companies.* If the annual report gives a breakdown of revenues from each of the industries in which the firm operates, then it may be possible to compare each of its divisions with averages for those industries. Most firms aspire to above average performance, so just comparing the firm with the industry average may not be adequate.

Inflation

Inflation has been a problem for quite some time, but accountants have had considerable difficulty in deciding the best way to take it into account. Inflation affects the real value of the firm's depreciation charges, its reported inventory costs, and its reported profits as well.

Accountants in the U.S. and Canada require firms to disclose some supplementary information in their annual reports, usually in footnotes to the financial statements. These footnotes may contain information about inflationary effects, but the financial statements themselves are still based on historical costs.

Accounting Practices

Different accounting practices followed by firms in the same industry can distort ratios. During inflationary periods, Last-In-First-Out (LIFO) inventory accounting produces a higher cost of goods sold and a lower end-of-the-reporting period inventory valuation than the First-In-First-Out (FIFO) method. Naturally, anything that affects the cost of goods sold will affect profits as well. Some firms use different depreciation methods for different types of assets. Even if they use the same depreciation method, they may use different economic lives for equipment that is otherwise comparable. These differences can be even more pronounced if the analyst is attempting to compare two firms from different countries.

Some firms capitalize their research and development expenses, while others charge R&D expenses against income each year.

Seasonality

Seasonality may also distort ratios. Inventories, receivables, and payables, are likely to have seasonal peaks and troughs. If ratios are calculated just before a peak, they will be very different from the same ratios calculated just after the peak. Thus, it is necessary to be sure that the end of the financial reporting period has not been astutely chosen so as to present an unusually good picture of the firm at the balance sheet date.

Common Size Analysis

Another widely used data analysis technique (not demonstrated here) is common size analysis. In a common size analysis all balance sheet items are divided by total assets and all income statement items are divided by net sales. Thus, common size statements show every item on the balance sheet as a percentage of assets and every item on the income statement as a percentage of sales. These percentages can then be compared to see if any items appear to be out of line. They are also useful in preparing forecasted, or pro forma, financial statements.

Interpretation of Financial Ratios

Financial ratios are usually analyzed in at least two ways: (1) by comparing the firm's ratios with industry or national average ratios or (2) by comparing trends in a firm's ratios over a period of several years.

Industry comparisons indicate how the firm compares with similar firms in the same industry.

Trend analysis shows whether the firm's ratios are improving or deteriorating, and this can be very important in attempting to forecast the future. Most analysts want to analyze trends in either ratios or percentages over a five to seven year period if the appropriate information is available. For many young and small firms just beginning their business life, such trend analyses are not possible. However, the necessary data will be available for established firms. To analyze a trend, it is frequently useful to have the spreadsheet draw a graph.

Trend analyses are useful for showing what has been happening to each ratio or percentage over time.

Data for comparative purposes can be obtained from industry trade associations, from financial reporting firms such as Dun and Bradstreet or Standard and Poor's, or from other sources. Some information is available on diskettes or magnetic tape or from on-line information utilities. Thus, easy access to up-to-date data is possible both for lenders and investors. Naturally, if comparisons are desired the ratios must be calculated in the same way.

Gandalf Example

An example financial data analysis based on data found in the 1990 Gandalf Technologies annual report demonstrates how the RATIOS worksheet can be used. Note that the currency format was not used in Figure 6-1 because of space limitations and that negative numbers are shown with a minus sign rather than with parentheses, which would normally appear. The RATIOS worksheet, combined with the Gandalf data in Figure 6-1, is called GANDALF.

Liquidity Ratios

The current ratio is displayed in cells C31, D31, and E31 of Figure 6-1. Although it is normally expected to be at least 2 to 1, data for all Canadian corporations indicates that in Canada the current ratio has been only 1.13. Gandalf's current ratio has declined from 2.70 to 2.03 to 1.87 during the three years for which we have data. The trend is definitely not good, but the absolute value of the ratio is acceptable since it is above the Canadian average, although slightly lower than the overall benchmark. Gandalf has almost twice as much invested in current assets as it owes in current liabilities. Thus, there are not likely to be any immediate problems in paying its bills as they come due, although we have no information about the due dates of the current liabilities or the maturity dates of the short term deposits included in current assets.

Since some of the current assets are inventories and a firm could have difficulty selling its products in a recession, another ratio is also used for liquidity analysis. The quick ratio is displayed in cells C32, D32, and E32 of Figure 6-1. The quick ratio is normally expected to be at least 1 to 1 but was only 0.91 for all Canadian corporations. Gandalf's quick ratio has also declined from 1.88 to 1.34 to 1.25 but is still above the norm. It therefore appears that Gandalf is in a satisfactory liquidity position, although the trend is not good.

	A	B	C	D	E
1	GANDALF				E46
2	FINANCIAL RATIOS	ENTER	YEAR	YEAR	YEAR
3	BALANCE SHEET ITEMS	DATA?	1988	1989	1990
4	Cash and Marketable Securities	Yes	14943	3841	9584
5	Accounts Receivable	Yes	43008	44308	39436
6	Inventories	Yes	27977	28230	27237
7	Other Current Assets	Yes	6035	7228	5134
8	Total Current Assets	No	91963	83607	81391
9	Total Net Fixed Assets	Yes	30131	34138	31180
10	Other Assets	Yes	5872	15847	14980
11	Total Assets	No	127966	133592	127551
12	Total Current Liabilities	Yes	34013	41225	43414
13	Total Long Term Liabilities	Yes	6399	6145	6139
14	Total Liabilities	No	40412	47370	49553
15	Total Equity	No	87554	86222	77998
16	Total Liabilities and Equity	No	127966	133592	127551
17	INCOME STATEMENT ITEMS	No	1988	1989	1990
18	Sales	Yes	160648	167369	161656
19	- Cost of Sales	Yes	75448	82153	83224
20	Gross Profit	No	85200	85216	78432
21	+ Other Revenues	Yes	0	0	0
22	- Other Expenses except Interest	Yes	69702	79820	81922
23	Earnings before Interest & Taxes	No	15498	5396	-3490
24	- Interest	Yes	757	1740	2225
25	Earnings before Taxes	No	14741	3656	-5715
26	- Taxes	Yes	6271	1635	615
27	Earnings after Taxes	No	8470	2021	-6330
28	+ Depreciation	Yes	6129	6889	8477
29	Net Cash Flow	No	14599	8910	2147

30	LIQUIDITY RATIOS	No	1988	1989	1990
31	Current Ratio	No	2.70	2.03	1.87
32	Quick Ratio	No	1.88	1.34	1.25
33	MANAGEMENT RATIOS	No	1988	1989	1990
34	Inventory Utilization	No	5.74	5.93	5.94
35	Receivables Turnover	No	3.74	3.78	4.10
36	Average Collection Period	No	97.72	96.63	89.04
37	Fixed Asset Utilization	No	5.33	4.90	5.18
38	Total Asset Utilization	No	1.26	1.25	1.27
39	DEBT RATIOS	No	1988	1989	1990
40	Debt Ratio	No	0.32	0.35	0.39
41	Times Interest Earned	No	20.47	3.10	-1.57
42	PROFITABILITY RATIOS	No	1988	1989	1990
43	Margin	No	5.27%	1.21%	-3.92%
44	Return on Assets	No	6.62%	1.51%	-4.96%
45	Return on Equity	No	9.57%	2.34%	-8.12%
46	Earning Power	No	12.11%	4.04%	-2.74%

Figure 6-1 GANDALF Worksheet

Management Ratios

Management ratios indicate how well management is using the firm's assets (things they own). They also provide clues about the quality of the assets as well as probable cash flows.

The inventory utilization (turnover) ratio, which shows how fast products are being sold, is displayed in cells C34, D34, and E34 of Figure 6-1. Gandalf's inventory utilization has remained relatively stable in the last two years at 5.94, after a slight increase from 5.74 in 1988. The norm for this ratio depends on the industry. The average value for this ratio for all Canadian corporations was 5.53 times. Gandalf's ratio is not very much different from this national average. Thus, this ratio does not provide us with much information in this particular case. We do know, however, that a value of six for this ratio indicates that inventory turns over approximately every two months.

Receivables turnover is displayed in cells C35, D35, and E35 of Figure 6-1. The norm for this ratio depends on the industry and the credit terms of the firm. If the firm normally gives its customers thirty days to pay after billing them at the end of each month, accounts receivable would be outstanding for at least 45 days on average and the turnover should be 8.1 times. For all Canadian firms, receivables turnover was 5.7 times. Gandalf's ratio has improved from 3.74 to 3.78 to 4.10, but is below the Canadian average. This is not good.

The other similar ratio, average collection period, is displayed in cells C36, D36, and E36 of Figure 6-1. It is more directly comparable to the firm's credit terms (which we don't know), since both are stated in terms of the number of days. Gandalf's average collection period has declined from 97.72 to 96.63 to 89.04 days. Since the average collection period for all Canadian corporations was 64 days, Gandalf's performance does not look very good by this criterion, unless their credit terms are ninety days.

Fixed asset turnover is displayed in cells C37, D37, and E37 of Figure 6-1. Gandalf's ratio has fluctuated from 5.33 to 4.90 to 8.18 so there is no apparent trend. The average for all Canadian corporations was only 0.82. Thus, Gandalf is using its fixed assets much more effectively than the average Canadian firm. Although this ratio is above the norm and Gandalf's performance on this criterion is currently good, it may also mean that Gandalf will have to expand its productive capacity in future.

Total asset turnover is displayed in cells C38, D38, and E38 of Figure 6-1. Gandalf's ratio has remained stable around 1.26. Since the average for all Canadian corporations is only 0.5, Gandalf is doing much better than average.

Debt Ratios

The debt ratio is displayed in cells C40, D40, and E40 of Figure 6-1. This ratio indicates directly how much of the funds employed by the firm come from creditors. Gandalf's ratio has increased somewhat from 32% to 35% to 39%. The average for all Canadian corporations was nearly double, at 71%. Hence, Gandalf is certainly in good financial shape according to this criterion.

The times interest earned ratio, which measures ability to pay interest, is displayed in cells C41, D41, and E41 of Figure 6-1. The average for all Canadian corporations was 2.38. Gandalf's ratio declined from 20.47 in 1988 to 3.10 in 1989 to -1.57 in 1990. Both the low absolute value (less than 6 is considered dangerous) of this ratio and its trend will cause creditors concern.

Profitability Ratios

The margin ratio is displayed in cells C43, D43, and E43 of Figure 6-1. The average for all Canadian corporations is 6%. Gandalf's ratio has fallen from the already low 5.27% in 1988 to only 1.21% in 1989 to -3.92% in 1990, as a result of the loss in that year. The trend is definitely bad.

Return on assets is displayed in cells C44, D44, and E44 of Figure 6-1. Gandalf's ratio has declined from 6.62% in 1988 to 1.51% in 1989 to -4.96% in 1990. The trend is not good. However, the 1988 ratio was more than double the average of only 3% for all Canadian corporations.

Return on equity is displayed in cells C45, D45, and E45 of Figure 6-1. Gandalf's ratio has declined from 9.67% in 1988 to 2.34% in 1989 to -8.12% in 1990. The average for all Canadian corporations was 11%. Thus, Gandalf was already below average in 1988. Its performance in 1990 was just awful.

Earning Power is displayed in cells C46, D46, and E46 of Figure 6-1. Gandalf's ratio has declined from 12.11% in 1988 to 4.04% in 1989 to -2.74% in 1990.

Ratio Analysis

Can we conclude that Gandalf is about to go bankrupt, given the very bad ratios in 1990? The answer is no, because 1990 may be an abnormal year. The 1990 loss was due to the recession which started then and continued into 1991. A firm in weaker financial shape than Gandalf might not be able to weather such a storm, but Gandalf can probably continue for at least three years. Sales are stable, inventory and receivables are turning over normally, there is ample cash, plus some short-term debt capacity (ability to borrow). Profitability is too low, and obviously cannot continue to be negative for very long, but research and development is continuing. If the firm could reduce its costs or increase sales, profits could rebound.

Formulae for RATIOS Worksheet

Figure 6-2 shows the first three data columns of the RATIOS worksheet. Not shown are columns A and B which merely identify the data inputs and were shown in Figure 6-1. Column B also serves as a reminder that data must be entered in rows 4 to 28 of columns C, D and E, wherever the word **Yes** appears in column B. Except for Row 3, these are the only rows in which data must be entered, since everything else is calculated by the model. Data for the years (1988, 1989, and 1990 for Gandalf) must be entered in Row 3. Cell C8, for example, contains a formula to add the numbers in cells C4, C5, C6, and C7. The formula in cell C11 is another way to add the numbers in cells C8, C9, and C10. Most of the other formulae are for simple arithmetical operations.

We should point out that since Gandalf is a Canadian firm, 365 days is used in cells C36 through E36 to calculate the average collection period. This is typical Canadian practice. However, you may want to change this to 360 days if you analyze a U.S. firm, as 360 is still typically used in the U.S.

Once the formulae have been entered in column C, they can be copied into the other columns by using the `/` `C` command. This command will cause the computer to make another cell, range of cells, row, or column just like the one being copied. Thus, once the formulae are entered in column C, columns D and E can be copied by the computer. As many more columns as are desired can be easily created this way.

The @IF and @COUNT functions could be added to the RATIOS worksheet to prevent the **ERR** or **ERROR** message appearing in the ratio cells before data is entered. Similarly, these functions could prevent $0 from appearing in balance sheet and income statement summations before data is entered. The WACC worksheet in Chapter 4 provided a demonstration of how to use these functions to blank the screen display.

	C	D	E
8	@SUM(C$4..C$7)	@SUM(D$4..D$7)	@SUM(E$4..E$7)
11	+C$8+C$9+C$10	+D$8+D$9+D$10	+E$8+E$9+E$10
14	+C$12+C$13	+D$12+D$13	+E$12+E$13
15	+C$11-C$14	+D$11-D$14	+E$11-E$14
16	+C$14+C$15	+D$14+D$15	+E$14+E$15
17	+C$3	+D$3	+E$3
20	+C$18-C$19	+D$18-D$19	+E$18-E$19
23	+C$20+C$21-C$22	+D$20+D$21-D$22	+E$20+E$21-E$22
25	+C$23-C$24	+D$23-D$24	+E$23-E$24
27	+C$25-C$26	+D$25-D$26	+E$25-E$26
29	+C$27+C$28	+D$27+D$28	+E$27+E$28
30	+C$3	+D$3	+E$3
31	+C$8/C$12	+D$8/D$12	+E$8/E$12
32	(C$8-C$6)/C$12	(D$8-D$6)/D$12	(E$8-E$6)/E$12
34	+C$18/C$6	+D$18/D$6	+E$18/E$6
35	+C$18/C$5	+D$18/D$5	+E$18/E$5
36	+C$5/(C$18/365)	+D$5/(D$18/365)	+E$5/(E$18/365)
37	+C$18/C$9	+D$18/D$9	+E$18/E$9
38	+C$18/C$11	+D$18/D$11	+E$18/E$11
40	+C$14/C$11	+D$14/D$11	+E$14/E$11
41	+C$23/C$24	+D$23/D$24	+E$23/E$24
43	+C$27/C$18	+D$27/D$18	+E$27/E$18
44	+C$27/C$11	+D$27/D$11	+E$27/E$11
45	+C$27/C$15	+D$27/D$15	+E$27/E$15
46	+C$23/C$11	+D$23/D$11	+E$23/E$11

Figure 6-2 Formulae for RATIOS and GANDALF Worksheet

Appendix A
SPREADSHEET COMMANDS

Spreadsheet software can perform a large number of commands. Commands are instructions to do certain things. These commands usually can be issued only when the program is in the READY mode. This appendix discusses some of the most often used commands which are common to the most popular spreadsheets.

To issue a command, depress the slash ⌐/⌐ key. The main menu will appear in the control panel and the mode indicator will say **MENU**.

Select the command by typing its first letter. Alternatively, you can select a command by moving the cursor until the desired command is highlighted and then depressing the (Enter) key.

The *main* menu commands are: **Worksheet, Range, Copy, Move, File, Print, Graph, Data, System, Add-In,** and **Quit.**

Except for **Quit,** each of these commands has submenus, and many of the submenus have submenus of their own.

Detailed comments on some of the more important items are included below.

Worksheet

The **W**orksheet command controls the appearance of the entire worksheet. The submenu commands are: **Global, Insert, Delete, Column, Erase, Titles, Window, Status, Page,** and **Learn.**

Global

The **G**lobal subcommand affects the entire worksheet. This command has submenus for **Format, Label-prefix, Column-width, Recalculation, Protection, Default,** and **Zero.**

Format sets the way numeric values appear in the entire worksheet.

Label-prefix sets the alignment of labels at the left or right margins of the columns, or centers them.

Column-width sets the width of all the columns in the worksheet to the same width.

Recalculation sets when the worksheet is to be recalculated, either automatically or manually (if you have lots of data); or sets how the recalculation will be effected, columnwise or rowwise.

Protection prevents unwanted changes from being made or allows changes. The submenu choices are **Enable** and **Disable.**

Default sets or lets you see what type of printer is connected, changes the margins, or specifies a printer setup string.

Zero suppresses or displays zeros on the screen.

Insert

The Insert subcommand adds columns or rows to the worksheet. Position the cursor in the column (row) where you want the new blank column (row) to be inserted. When you insert, the existing columns (rows) move over (down) to make room for the new ones. Formulae are adjusted automatically, so they continue to refer to the same data. If you insert into a named range, the size of the range increases.

Delete

The Delete subcommand deletes (removes) columns or rows from the worksheet. Delete works just like Insert but in the opposite direction. Ranges become smaller as a result of deletions. However, if you delete the corner of a range, the range will be invalidated. Formulae which refer to a deleted cell or range will show **ERR** for error.

Column

The Column subcommand changes the width of a column, and can hide or display it as well. It operates only on a specific column or range of columns. The submenus are: **Set-Width, Reset-Width, Hide, Display,** and **Column-Range**.

Set-Width changes the width of the current column, overriding the global setting.

Reset-Width resets the current column width to the global setting.

Hide can hide one or more columns.

Display can redisplay one or more previously hidden columns.

Column-Range changes the width of a range of columns.

Erase

The Erase subcommand erases the currently displayed worksheet from the screen and returns you to a blank worksheet.

Titles

The Titles subcommand freezes the top or side borders, or both of them, so you can still see them when you scroll to another portion of the worksheet.

Window

The Window subcommand divides the video screen into two windows, either vertically or horizontally. This is another way to see two parts of a worksheet simultaneously.

Status

The Status subcommand displays worksheet settings information including memory use, the global settings, and the hardware.

Page

The Page subcommand inserts a page-break into the worksheet so output will be one more than one printed page.

Learn

The Learn subcommand records keystrokes in worksheets.

Range

The Range command affects the display or elimination of a cell or range of cells. The range command has subcommands and several of these subcommands have their own subcommands as well. The subcommands are: **Format, Label, Erase, Name, Justify, Protect, Unprotect, Input, Value, Transpose,** and **Search.**

Format

There are numeric as well as non-numeric formats. The numeric Format subcommand effects the appearance of numbers in the worksheet, but it does not change their actual values in memory. The format chosen here overrides the default format. The formats are: **Fixed, Scientific, Currency, Comma, General, +/-, Percent, Date, Time, Text, Hidden,** and **Reset.**

Fixed format provides a constant number of decimal places with leading zero integers.

Scientific format can specify the number of decimal places in the multiplier; the exponent can be in the range of -99 to +99.

Currency is the format most often used in engineering economics. This format inserts a currency sign before or after the entry as desired, and also inserts a separator between thousands. Thus, the $ goes in front of the entry, and commas separate thousands. Negative values are enclosed in parentheses.

Comma format is similar to currency but without the currency sign.

General format suppresses trailing zeros. It is the default format.

+/- format makes horizontal bar charts, using **+** for positive values, **-** for negative values, and **.** for values between -1 and +1, including zero.

Percent format displays the value times 100 followed by the % sign.

Date format allows choice between American and International formats with either month or day first.

Time format allows choice between American and International formats, either 12 or 24 hours.

Text format replaces numeric values with the cell's formula.

Hidden format hides the contents of a specified range so the screen won't display it.

Reset format restores the default numeric format.

Label

The Label subcommand aligns labels in a range of cells at the left or right edge or in the center of each cell.

Erase

The Erase subcommand removes the contents of a cell or a range of cells.

Name

The Name subcommand names a range or redefines the range of cells to which a range name refers.

Justify

The Justify subcommand treats a continuous column of text as a paragraph, rearranging the words so that none of the lines is longer than a specified width. This can be useful for small amounts of text processing. *However, do not use this command if any cells in the range are protected!*

Protect

The Protect subcommand prevents changes and deletions to a range of cells when the global protection for the worksheet is enabled and the unprotect command has been used. It does *not* protect if global protection is disabled.

Unprotect

The Unprotect subcommand allows changes to a range of cells when global protection for the worksheet is enabled. The protect command above is then used to toggle back to protection for an individual cell or a range of cells that may be smaller than the original range.

Input

The Input subcommand limits movement of the cursor to the unprotected cells within a specified range. This command can be used to set up fill-in-the-blanks entry forms for data entry purposes, as in the case of combining reports.

Value

The Value subcommand copies the displayed values instead of the formulae from one cell to another. *Use this command only if the range to which you are moving the values is empty, because it will overwrite whatever may be there!*

Transpose

The Transpose subcommand copies and rearranges data from a column to a row or vice-versa. The original range remains as it is, only the copy is rearranged.

Search

The Search subcommand searches for specified character strings (such as XYZ) within a specified range. It is not case sensitive. That is, XYZ is equivalent to XYz, Xyz, and so forth.

Copy

The Copy command reproduces entries from one cell to another. Exact duplicates of labels and numbers are made. *However, for the formulae, spreadsheet software may or may not adjust the cell addresses, depending on the kind of cell address used.* There are three kinds of cell addresses: Absolute, Relative, and Mixed.

When cell addresses are entered into formulae, spreadsheets normally record only the address of the inputs in relation to the cell where the formula is located.

For example, if a cell D8 contained the formula +D4+D5, the spreadsheet would interpret this as being an instruction to add the value found 4 rows up to the value found 3 rows up, and put the sum in D8. If this is not what you want a formula to do, then you can use an absolute address, such as $D$4, to indicate that a value from a particular location is to be used. Preceding a range name by a $ makes it absolute. If you want the row to be constant, while the column varies, use a mixed address, such as D$4.

Move

The **Move** command moves entries from one location to another. This is similar to copy, but the entry in the old location is erased. Formulae throughout the worksheet are adjusted as necessary to account for data that has been moved, irrespective of whether the references are relative, mixed, or absolute. *You should move only to an empty portion of the worksheet.*

File

The **File** command allows you to retrieve and save files. File names can be eight characters long. There are three types of files: Worksheet, Print, and Graph. These files are identified by their three-character extensions. Worksheets are identified by .WKS, .WK1, or similar designation. Print files have the .PRN extension. Graph files have the .PIC extension. An Other category allows you to enter your own extensions when you save a file. The subcommands are: **Retrieve, Save, Combine, Xtract, Erase, List, Import, Directory**, and **Admin.**

Retrieve

The **R**etrieve subcommand loads a file from a disk into the computer memory and displays it on the screen as well. Use this command to retrieve the files from the diskette which accompanies this book. You will be prompted by the software, which will display some of the files available on the current default data drive. Use the cursor to highlight the file you want, then press the Enter key. If you want to retrieve something from a drive that is not the default drive, you must first tell the computer where to look by specifying the drive with the appropriate designator, such as B:. For more information on retrieving a worksheet, see Chapter 1.

Save

The **S**ave subcommand saves the current worksheet and any associated settings to disk. Use this command often, so you will have a recent version of your work in case of power failure or other calamity, such as a network failure. **It is recommended that you save your work every 5 to 15 minutes.** It is a good idea to use two back-up diskettes, alternating them as you go. That is, save on the first one after 5 minutes, then use the second one after another 5 minutes, then use the first one again after five more minutes. To save on a drive which is not the default, you must precede the file name with the appropriate drive identifier, such as B:. For more information on saving a worksheet, see Chapter 1. The subcommands are **Cancel** and **Replace**.

Cancel Use **C**ancel to return to the worksheet without saving.

Replace Use **R**eplace to update a file. If your diskette does not have sufficient room for the file you are trying to save, the spreadsheet will give you an error message. If this happens, press (Esc) , insert another formatted data disk, and try to save again. **Thus, it is important to have a sufficient quantity of formatted data diskettes before beginning a spreadsheet software session.**

Combine

The Combine subcommand incorporates all or part of a worksheet into the current worksheet at the location of the cursor. The incoming worksheet will fill in the area to the right and below the cursor position. You should make sure this area is empty before combining. The submenu choices are: **Copy, Add,** and **Subtract.**

Copy writes over whatever might be in the current worksheet.

Add will add the incoming worksheet into the current worksheet.

Subtract will erase (subtract) a range from the current worksheet.

Xtract

The **X**tract subcommand extracts and saves a portion of the current worksheet as a separate worksheet. Use this command to split a large worksheet into smaller parts or to use a part of a worksheet in another worksheet. *When extracting, make sure that any formulae within the range to be extracted do not reference cells outside the range.*

Erase

The Erase subcommand removes (deletes) files from the disk. The submenu choices are: **Worksheet, Print, Graph,** and **Other.** The first three can be used to display filenames of that type. Then you can select the file to be erased by moving the cursor to highlight the filename. Press (Enter) and then answer **Yes** or **No** from the menu.

List

The List subcommand lists the files of a particular type. It uses the same submenu choices as the **Erase** command: **Worksheet, Print, Graph** and **Other.**

Import

The Import subcommand copies a print file from the current directory into the worksheet at the location of the cursor. Only files that have the .PRN extension can be imported, so you may have to rename word processor files. ASCII files must be less than the maximum size of 240 by 8192 characters. (ASCII means American Standard Code for Information Interchange. It is the standard for microcomputers.) Some word processors use special characters for formatting. These special characters will cause unpredictable problems in most spreadsheets, so they should be removed before importing the file.

Directory

The Directory subcommand changes the current directory to a new one specified by the user.

Admin

The Admin subcommand has three submenu choices. **Link-Refresh** recalculates formulae in the current worksheet that include references to files on disk. **Reservation** allows worksheet files to be shared on networks. **Table** creates a table of information about files on disk.

Print

The Print command allows you to make printouts. The main submenu has only two choices: **Printer** or **File**. The Printer subcommand prints directly to a printer. The File subcommand prints the file to disk. Printing to disk gives the file the .PRN extension and allows the file to be printed from DOS.

Printer and File

The subcommand menu choices for both Printer and File are: **Range, Line, Page, Options, Clear, Align, Go**, and **Quit.**

Range lets you specify the range you want printed or stored on disk.

Line lets you advance the paper in the printer by one line.

Page lets you advance the paper in the printer to the next page. You will normally need to use this command after every **Go** and before **Quit** to make sure that the paper is advanced to the top of the page so it will be ready for the next printout job. If you are using a laser printer, you must use **Page** command. Some printers require the **Page** command before the **Go** command.

Options submenu choices are: **Header, Footer, Margins, Borders, Setup, Page-length, Other**, and **Quit**.

▪ *Header* prints one line of text just below the top margin of every page. You can type up to 240 characters but you will be limited by the size of the paper and the margins. Use a ⒡ to generate sequential page numbers on the pages. Use an ⒜ to produce the date.

▪ *Footer* prints one line of text just above the bottom margin. It is equivalent to header.

▪ *Margins* sets the margins for the printed page and overrides the default margins. The submenu has four choices: **Left, Right, Top**, and **Bottom**.

▪ *Borders* prints specified columns or rows on each page, above or to the left of the range which is being printed. The submenu has 2 choices: **Rows** and **Columns**.

▪ *Setup* specifies the font size and style for the printer. To change the setup string, hit (Esc) first. Then type the setup string, which is usually a three or four digit code, preceded by a backslash ⒩ .

▪ *Page-length* indicates the number of printed lines that fit on the page. Press (Enter) to use the current length, or type a number between 10 and 100; then press (Enter) .

▪ *Other* allows you to choose from the submenu's four choices: **As- displayed, Cell-formulas, Formatted**, and **Unformatted**.

▫ **As-displayed** will print the worksheet as it appears on the screen.

▫ **Cell-formulas** will print the contents of each filled cell rather than the displayed values.

▫ **Formatted** will restore any page breaks, headers, and footers after you have chosen **Unformatted**.

▫ **Unformatted** prints without page breaks, headers, or footers. This can be useful if you want to print a very full page and get everything on one sheet of paper.

■ *Quit* returns to the previous menu.

Clear returns to As-displayed and/or restores other settings to the default values. The submenu has four choices: **All, Range, Border,** and **Format**.

■ *All* cancels the current print range, clears all borders, headers, and footer, and resets all formats and options to the default settings.

■ *Range* cancels the current print range.

■ *Borders* clears all borders.

■ *Format* resets margins, page length, and the setup string to the default values.

Align tells the printer that you have positioned the paper in the printer to the top of the page.

Go starts the printing process if a range has been specified.

Quit tells the computer that you are finished with the **Print** command. You must select **Quit** after you have used **Go**, or the process will be unfinished. Normally you will need to use **Page** between **Go** and **Quit**. If you are using a laser printer, you probably need to use **Page** (some lasers require it) before **Go** instead of afterwards.

Graph

The **Graph** command allows you to create and display graphs from data contained in worksheets. The submenu choices are: **Type, XABCDEF, Reset, View, Save, Options, Name,** and **Quit**.

Type

The Type subcommand allows you to choose one of the graph types. They are **Line, Bar, XY, Stacked-bar**, and **Pie**.

Each graph can contain six variables. Each of the variables will be represented by a different symbol, chosen by the spreadsheet.

Line graphs represent each value with a point at an appropriate spot on the graph. The various points are connected by a line.

Bar graphs represent each value with a bar of varying height.

XY graphs pair values, with X on the horizontal axis and Y on the vertical.

Stacked-bar graphs display corresponding values from each relevant data range stacked one atop the other.

Pie charts compare parts to the whole, so each value is a wedge shaped piece of the pie. You can explode one or more of the wedges for emphasis.

XABCDEF

This subcommand allows you to set the data ranges, with X for the X data range, A for the first, B for the second, etc.

Reset

The **Reset** subcommand cancels graph or range settings. The submenu choices are **X, A, B, C, D, E, F**, and **Quit**. The A to F identify the ranges where the six variables are located. X identifies the location of the label entries. **Quit** returns you to the previous menu.

View

The **View** subcommand displays the graph on the screen. The [F10] key does the same thing from the READY mode.

Save

The Save subcommand saves the graph as a .PIC file.

Options

The Options subcommand leads you to the submenu choices of **Legend, Format, Titles, Grid, Scale, Color, B&W, Data-Labels,** and **Quit.** Several of the submenus have submenus.

Legend adds a legend below the graph to explain the symbols, colors, or crosshatching.

Format controls the appearance of line and XY graphs. The submenu has 8 choices: **Graph, A, B, C, D, E, F,** and **Quit.**

■ *Graph* controls the format of the entire graph.

■ *A* through *F* control the format of each data range. The submenu has 4 choices: **Lines, Symbols, Both** and **Neither.**

□ **Lines** connects each range with a straight line.

□ **Symbols** displays each point with the same symbol.

□ **Both** sets both lines and symbols.

□ **Neither** deletes both lines and symbols, and you must then specify data labels to make the points or lines visible.

□ **Quit** returns to the previous menu.

Titles assigns a title to the entire graph or to the axes, or all of them.

■ *First* and *Second* refer to the main and sub titles for the graph. These titles are centered at the top of the graph. The main title will be printed in a larger font than the sub title.

■ The *X-Axis* title appears below the horizontal axis.

■ The *Y-Axis* title appears to the left of the vertical axis.

Grid adds or removes grid lines on the display, except for pie charts. **Horizontal** or **Vertical** or **Both** may be selected. **Clear** erases all the grid lines.

Scale sets the numeric scales for the X and Y axes. Skip allows only selected items to be printed, so graphs are not so crowded. For example, a skip factor of 10 will cause only the first, eleventh, and every other tenth item to print. For both the X and Y scales, the submenu allows you to choose from among **Automatic, Manual, Lower, Upper, Format, Indicator**, and **Quit**.

■ *Automatic* displays all the data points using scale limits which fill up the screen. This is the default setting and you should probably try it first when you make a graph.

■ *Manual* makes you specify the lower and upper limits.

■ *Lower* determines the lower scale limit.

■ *Upper* determines the upper scale limit.

■ *Format* allows you to choose any one of the available formats.

■ *Indicator* allows you to choose whether or not to display the scale indicators.

■ *Quit* returns to the previous menu.

Color allows you to choose color settings, if your monitor can display color graphs.

B&W allows you to choose monochrome crosshatchings.

Data-Labels labels the data points within a range. On line and XY graphs, labels may be centered, left or right aligned, or put below their data points. On bar and stacked-bar charts, labels are centered above positive bars and below negative bars. On pie charts, labels are not displayed.

Quit returns to the previous menu.

Name

The **Name** subcommand allows you to choose from a submenu of four choices: **Use, Create, Delete**, and **Reset**.

Use makes the named set of graph settings current and draws the graph.

Create saves the current graph settings under a graph name. This command can be used to save more than one graph with a worksheet with different names or to shift from one graph to another.

Delete erases one set of graph specifications. Use this to eliminate unwanted graphs or to free a graph name for reuse with new data.

Reset erases all named graphs.

Quit

The **Quit** subcommand returns to the previous menu.

Data

The **Data** command allows you to manipulate a worksheet database. A worksheet database is a worksheet range that contains data. All entries in a single row constitute a record, which is a collection of information about one item in a database. Each record consists of categories called fields. A database can have up to 256 fields. Each of the columns in a database is a field. Fields may contain either labels or numeric values, but not both in the same field. Each cell in the first row of the database can contain a label which will identify the data in the field. This row (or rows), however, must be excluded from the range of the main menu commands discussed below. The other rows contain the records. Blank rows are not allowed. The main menu commands are: **Fill, Table, Sort, Query, Distribution, Matrix, Regression**, and **Parse**.

Fill

The **Fill** subcommand enters an ascending or descending sequence of numbers into a specified range of cells.

Table

The Table subcommand allows you to try out different values in a formula, so it is very useful for sensitivity analyses. The submenu choices are 1, 2, and Reset. The first two let you choose to change 1 variable or 2 for each iteration, and reset returns to the default setting.

Sort

The **S**ort subcommand rearranges data as specified, in ascending or descending order.

The submenu choices are **Data-Range, Primary-Key, Secondary-Key, Reset, Go**, and **Quit.**

Data-Range indicates which records are to be sorted.

Primary-Key indicates the field to be used to determine the new order of the records.

Secondary-Key indicates the field to be used to break ties when two or more records have the same entries in the primary-key field.

Reset returns everything to the default settings.

Go is the signal to sort.

Quit returns to READY mode.

Query

The **Q**uery subcommand lets you search for a particular item in a database as well as copy, extract, or remove selected records. The submenu choices are: **Input, Criterion, Output, Find, Extract, Unique, Delete, Reset**, and **Quit.**

Input indicates where to search.

Criterion indicates how to search.

Output indicates where to put what is found.

Find locates records.

Extract copies records.

Unique eliminates any duplicates.

Delete erases records.

Reset cancels input, criterion, and output ranges.

Quit returns to READY mode.

Distribution

The **D**istribution subcommand creates a frequency distribution of the values in a specified range.

Matrix

The **M**atrix subcommand multiplies and inverts matrices, formed by rows and columns, up to size 90 by 90. Only square matrices can be inverted. When multiplying matrices, there must be the same number of columns in the first range as there are rows in the second range.

Regression

The **R**egression subcommand computes the coefficient values and constant for a formula that relates one or more ranges of independent variables to a range of dependent variables. It also indicates the statistical accuracy of these values. You can specify up to 16 independent variables for multiple regression. In the results, the X coefficients are slopes and the Y intercept is a constant. The X and Y ranges must have the same number of rows. The submenu choices are: **X-Range, Y-Range, Output-Range, Intercept, Reset, Go,** and **Quit.** Most of these should be self-explanatory. **Intercept** allows you to choose between computing the Y intercept and forcing it to zero. The default is to compute it. See Chapter 3 for a discussion of the REGRESS worksheet which illustrates most of these submenu choices.

Parse

The **P**arse subcommand converts a column of long labels into several columns. Use this command to convert an ASCII file you have imported into a spreadsheet. The submenu choices are: **Format-Line, Input-Column, Output-Range, Reset, Go**, and **Quit**.

System

The **S**ystem command lets you temporarily leave your worksheet, use DOS, and return to the worksheet without restarting.

Add-In

The **A**dd-In command attaches, detaches, invokes, or clears add-in programs.

Quit

The **Q**uit command ends the spreadsheet session and returns to DOS.

Appendix B

SPREADSHEET FUNCTIONS

Spreadsheet functions are built-in formulae which perform special types of calculations that may need to be done frequently. The functions make it unnecessary for the user to reinvent the wheel for these calculations.

Functions usually begin with the **@** character. (SuperCalc will accept the @ character but does not require it.) An example function is @SUM(K5..K10). Thus, **@SUM** is the function name, and **(K5..K10)** is the argument of the function. Functions are written without any spaces between the function name and the arguments. Names are usually typed in uppercase letters. Arguments can be of three types: numeric, range, and string values.

The most useful functions for engineering economists are the financial and statistical functions. There are also mathematical, logical, special, string, as well as date and time functions. For more information on specific spreadsheet functions consult the appropriate Reference Manual.

Some of the typical functions common to all the major spreadsheets are discussed below. Some minor differences between spreadsheets are: Quattro calls ranges blocks, while SuperCalc and VP Planner use present value for principal, and VP Planner uses Estimate instead of Guess in some functions.

Financial Functions

The financial functions make calculations concerning loans, annuities, etc. Interest rates can be entered either as percentages or as decimal fractions. Thus, **15.5 percent** can be entered either as **15.5%** or as **.155**. It is recommended that you use the decimal format, **.155**. The term and the interest rate must always be in the same units of time. For example, suppose you want to calculate a monthly payment when the nominal or stated interest rate is given in years *and compounding is monthly*. Divide the annual nominal rate by 12 to get the effective monthly periodic interest rate, and multiply the term in years by 12 to get the number of monthly payment periods. It is advisable to format cells where monetary results are expected as *Currency*, while cells which are to receive percentage results should be formatted as *Percent*.

@CTERM(INT,FV,PV) This function returns the number of periods for an investment of present worth PV to grow to a future worth FV at the periodic interest (compounding) rate INT.

@FV(PMT,INT,TERM) This function returns the future worth of a series of equal payments PMT earning periodic interest at the rate INT over the number of payment periods TERM. In ANSI notation, this function returns a future worth F that is the product of a uniform series amount A times a uniform series compound amount factor (F/A,i,N).

@IRR(GUESS,RANGE) This function returns the internal rate of return for the series of cash flows in RANGE, based on the GUESS interest rate. The function iterates until it finds the IRR.

@NPV(INT,RANGE) This function returns the present worth of the series of cash flows in RANGE, discounted at the interest rate INT. *This function assumes that the first cash flow in the specified range is one period from today* and that the last cash flow is N periods from today, where N is the number of cells in the range.

@PMT(PRIN,INT,TERM) This function returns the amount of the periodic payment required to pay off the principal amount PRIN at the periodic interest rate INT over the number of periods TERM. In ANSI notation, this function returns a uniform series amount A that is the product of a present worth P times a capital recovery factor (A/P,i,N).

@PV(PMT,INT,TERM) This function returns the present worth of a series of equal payments PMT discounted at the periodic rate INT over the number of payment periods TERM. In ANSI notation, this function returns a present worth P that is the product of a uniform series amount A times the uniform series present worth factor (P/A,i,N).

@RATE(FV,PV,TERM) This function returns the periodic interest (compounding) rate necessary for a present worth PV to grow to a future worth FV over the number of payment periods TERM.

Mathematical and Statistical Functions

@AVG(LIST) This function returns the average of the values in LIST.

@EXP(X) This function raises the number e to the xth power.

@LN(X) This function calculates the natural logarithm of X.

@LOG(X) This function calculates the logarithm of X.

@RAND This function returns a random number between 0 and 1.

@ROUND(X,N) This function returns X rounded to N places.

@SQRT(X) This function returns the positive square root of X.

@STD(LIST) This function returns the standard deviation of the values in LIST.

@SUM(LIST) This function returns the sum of the values in LIST.

@VAR(LIST) This function returns the variance of the values in LIST.

Special Functions

@HLOOKUP(X,RANGE,ROW) This function performs a horizontal table lookup. It returns the contents of the cell that is the specified number of ROWs below a reference cell in the top row of the RANGE. The reference cell is one cell to the left of the cell in the top row whose contents exceed X. X may be entered as either a formula, a cell reference, or a specific value.

For example, suppose @HLOOKUP($A3,GRADE,1) is entered in the target cell B3, where cell A3 contains a student's numerical grade of 82.5 in engineering economics and GRADE is defined by the slash **R**ange **N**ame **C**reate command ⟨/⟩⟨R⟩⟨N⟩⟨C⟩ as A1.F2. The contents of the range GRADE are contained in cells A1 through F2 in Figure B-1.

First @HLOOKUP($A3,RANGE,1) will select cell D1 as the reference cell in the range GRADE because cell E1 is the first cell whose contents of 89.5 exceed 82.5. Then the function will return the contents of cell D2 to the target cell B3 because D2 is the specified number of rows 1 below the reference cell D1. That is, the student's letter grade of B will appear in cell B3 as shown in Figure B-1. Grades of additional students can be entered in column A, starting with row 4. The @HLOOKUP formula in cell B3 can be copied to as many cells in column B as necessary, starting with cell B4.

	A	B	C	D	E	F
1	0	59.5	69.5	79.5	89.5	100.1
2	F	D	C	B	A	
3	82.5	B				

Figure B-1 An Example @HLOOKUP(X,RANGE,ROW)

@VLOOKUP(X,RANGE,COLUMN) This function performs a vertical table lookup. It returns the contents of the cell that is the specified number of COLUMNs to the right of a reference cell in the first column of the RANGE. The reference cell is one cell above the cell in the first column whose contents exceed X. X may be entered as either a formula, a cell reference, or a specific value. An example of the @VLOOKUP function is in the NPWTAX worksheet in cell B16 in Figure 5-2A in Chapter 5.

Appendix C

SPREADSHEET MACROS

Spreadsheet macros are generally considered to be an advanced topic because it is possible to use spreadsheets without them. However, they are a means of greatly increasing productivity, so you should be aware of them.

Macros are lines of code entered as labels in an out of the way part of the spreadsheet. To use the fast method for running a macro, any one letter of the alphabet may be used to name a macro. The letter must be preceded by a backslash. For example, a macro might be named \R. The macro name is located in the cell to the left of the macro instructions. The macro name must entered in the appropriate cell as a label, *viz.,* `' \ R ⏎` . Once the macro \R is created and its range is named, it can be invoked by holding down the MACRO key labelled `Alt` and then pressing the letter `R` .

Obviously, the longer the macro, the more keystrokes are saved. This reduces the likelihood of errors. Since computers are also much faster at interpreting the instructions than people are at pressing the keys, macros will speed up any keystroking operation.

Almost all possible macros are just sequences of keystrokes grouped together to perform some operation. Thus, the macro can contain alphanumeric characters. There are also some special keys which must be enclosed in braces { }. For more information on these special keys consult the appropriate manuals.

The Regress Reset macro is shown in Figure C-1.

	C	H
1	\R	{GOTO}A1 ~
2		REGRESS ~
3		{GOTO}D1 ~
4		"R416 ~
5		/REA4.F9 ~
6		/REA11.R12 ~
7		/REA17.R416 ~

Figure C-1 The Regress Reset Macro

The Regress Reset Macro

The \R macro resets any REGRESS worksheet with data to the original state shown in Figure 3A-1 of Appendix 3A.

If you do not have the ESTSALES worksheet on your monitor, follow the instructions in Appendix 3A to create it.

Type ⑦ⓦⒼⓅⒹ to disable global protection.

Scroll to cell G1, type ⑦⟨\⟩Ⓡ→ and \R should appear.

You can now enter the codes for the macro.

In cell H1, type ⑦⟨{⟩ⒼⓄⓉⓄ⟨}⟩Ⓐ⓵~⬇ .

In cell H2, type ⑦ⓇⒺⒼⓇⒺⓈⓈ~⬇ . The codes in cells H1 and H2 will reset the label in cell A1 to REGRESS when the macro is invoked.

In cell H3, type ⑦⟨{⟩ⒼⓄⓉⓄ⟨}⟩Ⓓ⓵~⬇ .

In cell H4, type ⑦"Ⓡⓐ⓵⑥~⬇ . The codes in cells H3 and H4 will restore R416 as the lower print range specified in cell D1.

In cell H5, type `' / R E A 4 . F 9 ~ ↓` .

In cell H6, type `' / R E A 1 1 . R 1 2 ~ ↓` .

In cell H7, type `' / R E A 1 7 . R 4 1 6 ~ ⏎` . The codes in cells H5, H6, and H7 erase the relevant cells in the ranges A4..R13 and A17..R416. (Rows 10 and 13 have formulae that blank the screen when the ranges A4..F9 and A11..R12 are erased.) At this point the macro name and instructions should appear as in Figure C-1.

Name the macro by typing `/ R N C \ R H 1 . H 7 ⏎` .

Restore global protection by typing `/ W G P E` .

Run (or invoke) the macro by holding down the `Alt` key and pressing the `R` key. The screen should then look like the REGRESS worksheet shown in Figure 3A-1 of Appendix 3A.

You may wish to replace the original REGRESS worksheet on your template diskette with this macro version.

BIBLIOGRAPHY

Adamis, Eddie, *Command Performance Lotus 1-2-3* (Bellevue, Washington: Microsoft Press, 1986).

Anderson, David, Dennis J. Sweeney, and Thomas A. Williams, *Statistics for Business and Economics*, Fourth Edition (St. Paul: West Publishing Company, 1990).

Anderson, Dick and Bill Weil, *1-2-3 Tips, Tricks, and Traps*, Third Edition (Carmel, Indiana: Que Corporation, 1989).

Anderson, Dick and Bill Weil, *1-2-3 Database Techniques* (Carmel, Indiana: Que Corporation, 1989).

Au, Tung and Thomas P. Au, *Engineering Economics for Capital Investment Analysis* (Boston: Allyn and Bacon, 1983).

Blank, Leland T. and Anthony J. Tarquin, *Engineering Economy*, Third Edition (New York: McGraw-Hill, 1988).

Brigham, Eugene F., Alfred L. Kahl, William F. Rentz, and Louis C. Gapenski, *Canadian Financial Management*, Third Edition (Toronto: Holt, Rinehart and Winston of Canada, 1991).

Buck, James R., *Economic Risk Decisions in Engineering and Management* (Ames: Iowa State University Press, 1989).

Cobb, Douglas F., and Leith Anderson, *1-2-3 for Business* (Carmel, Indiana: Que Corporation, 1984).

Collier, Courtland and William B. Ledbetter, *Engineering Economics and Cost Analysis*, Second Edition (New York: HarperCollins, 1988).

Couper, James R., and William H. Rader, *Applied Finance and Economic Analysis for Scientists and Engineers* (New York: Van Nostrand, 1986).

Criner, E. A., *Successful Cost Reduction Programs for Engineers and Managers* (New York: Van Nostrand, 1984).

DeGarmo, Paul E., William G. Sullivan, and J. A. Bontadelli, *Engineering Economy*, Eighth Edition (New York: Macmillan, 1989).

Ewing, David P., *et. al.*, *Using 1-2-3, Special Edition* (Carmel, Indiana: Que Corporation, 1987).

Fleischer, Gerald A., *Engineering Economy: Capital Allocation Theory* (Monterey, California: Brooks/Cole, 1984).

Grant, Eugene L., William G. Ireson, and Richard S. Leavenworth, *Principles of Engineering Economy*, Eighth Edition (New York: Wiley, 1990).

Kleinfeld, Ira H., *Engineering and Managerial Economics* (New York: Holt, Rinehart and Winston, 1986).

Kurtz, Max, *Engineering Economics for Professional Engineer's Examinations*, Third Edition (New York: McGraw-Hill, 1985).

Lang, Hans J., *Cost Analysis for Capital Investment Decisions* (New York: Marcel Dekker, 1989).

McGuigan, James R. and R. Charles Moyer, *Managerial Economics*, Fifth Edition (St. Paul: West Publishing Company, 1989).

Moyer, R. Charles, James R. McGuigan, and William J. Kretlow, *Contemporary Financial Management*, Fourth Edition (St. Paul: West Publishing Company, 1990).

Nash, John C., *Compact Numerical Methods for Computers: Linear Algebra and Function Minimisation*, Second Edition (Bristol, U.K.: Adam Hilger, 1990).

Newnan, Donald G., *Engineering Economic Analysis*, Fourth Edition (San Jose, California: Engineering Press, 1990).

O'Leary, Timothy and Linda O'Leary, *The Student Edition of Lotus 1-2-3 Release 2.2* (Reading, Massachusetts: Addison Wesley, 1990).

Orvis, William J., *1-2-3 For Scientists & Engineers* (Alameda, California: Sybex, 1987).

Oulman, Charles S. and Motoko Y. Lee, *Macro Programming Using 1-2-3 for Engineers* (St. Paul: West Publishing Company, 1990).

Park, Chan S. and Gunter P. Sharp-Bette, *Advanced Engineering Economics* (New York: Wiley, 1990).

Person, Ron, *1-2-3 Business Formula Handbook* (Carmel, Indiana: Que Corporation, 1986).

Person, Ron, *et. al.*, *Using 1-2-3, Release 3 Edition* (Carmel, Indiana: Que Corporation, 1989).

Riggs, James L. and Thomas M. West, *Engineering Economics*, Third Edition (New York: McGraw-Hill, 1986).

Riggs, James L., William F. Rentz, Alfred L. Kahl, and Thomas M. West, *Engineering Economics, First Canadian Edition* (Toronto: McGraw-Hill Ryerson, 1986).

Sagman, Stephen W., *1-2-3 Graphics Techniques* (Carmel, Indiana: Que Corporation, 1990).

Smith, Gerald W., *Engineering Economy: Analysis of Capital Expenditures*, Fourth Edition (Ames: Iowa State University Press, 1987).

Sprague, J.C., and J. D. Whittaker, *Economic Analysis for Engineers and Managers* (Englewood Cliffs, N.J.: Prentice-Hall, 1986).

Thuesen, Gerald J. and Walter J. Fabrycky, *Engineering Economy*, Seventh Edition (Englewood Cliffs, N.J.: Prentice-Hall, 1989).

White, John A., Marvin H. Agee and Kenneth E. Case, *Principles of Engineering Economy*, Third Edition (New York: Wiley, 1989).

INDEX

Notes

Notes

Notes

Notes

Notes

Notes